BRICKWORK FOR APPRENTICES

4th edition

J. C. HODGE L.M.G.B.

Full Technological Certificate in Brickwork, City and Guilds of London Institute,
First-class Certificate, Building Construction and Sanitation, Carpenter's
Company, Former Senior Lecturer, Northern Polytechnic, London.

Revised by

R. J. BALDWIN P.P.G.B.

Construction Industry Consultant. Full Technological Certificate in Brickwork,
City and Guilds of London Institute. Higher National Certificate: Building.
Former Senior Lecturer, Willesden College of Technology, London.

ELSEVIER
BUTTERWORTH
HEINEMANN

AMSTERDAM · BOSTON · HEIDELBERG · LONDON · NEW YORK · OXFORD
PARIS · SAN DIEGO · SAN FRANCISCO · SINGAPORE · SYDNEY · TOKYO

Elsevier Butterworth-Heinemann
Linacre House, Jordan Hill, Oxford OX2 8DP
30 Corporate Drive, Burlington, MA 01803

First published 1944
Second edition 1960
Third edition 1971
Fourth edition 1993
Reprinted 2001, 2002 (twice), 2003, 2004
Transferred to digital printing 2004, 2005

British Library Cataloguing in Publication Data
A catalogue record for this book is available from the British Library

ISBN 0 340 55641 2

For information on all Elsevier Butterworth-Heinemann
publications visit our website at www.bh.com

Typeset in 10/11 pt Linotron Times by Wearset, Boldon, Tyne and Wear
Printed and bound by Lightning Source

Preface to fourth edition

It has been a privilege to assist J. C. in updating his book for a fourth edition. *Brickwork for Apprentices* has been the basic reference book on brickwork for generations of bricklayers, having been in continuous publication since 1944. Great care has been taken to retain the book's original intent. Namely, to give a clear understanding of craft techniques together with associated knowledge by using plenty of diagrams, on the basis that 'a good sketch is worth a thousand words'.

This remains the ideal reference book for all courses and schemes of training for brickwork and the craft skill of bricklaying, whether these are based at schools, training centres or Colleges of Technology.

The original title *Brickwork for Apprentices* has been deliberately retained for reasons given in Chapter 1.

The principle changes to this Fourth Edition include a new chapter devoted to bricks of special shapes and sizes, and another concerning external works brickwork. In addition, cavity wall construction has been transferred to its own chapter and grouped with the subject of brick cladding to framed buildings.

The principle deletion, after much heart searching, has been the finely detailed chapter about Gauged brickwork. The reasons for this are that genuine Red Rubber bricks are no longer available and the highly labour intensive art of rubbing bricks has been displaced by the availability of direct cutting to shape with masonry bench saws, where very fine jointed gauged arches are required.

The distinctive style and quality of John Hodge's own original drawings has been retained for this 4th Edition of 'Brickwork for Apprentices'

Acknowledgement _____

We are grateful to James Griffin for his enthusiastic interest throughout the updating of this fourth edition; and acknowledge his original suggestion for presentation of mortar selection information in Fig. 2.7, which was dubbed the 'Griffin Chart' during the review process.

JCH, RJB

Preface to first edition

Long experience in the teaching of Brickwork to craft students has led to the writing of this work. While there are many specialised books on the subject, I have felt that this small book may help the apprentice who wishes to have, in readily available form, information relating to the work he will have to perform and the knowledge he will require at the earlier stages of his career.

Although I have written mainly for apprentices, particularly those taking part-time courses in technical colleges, I hope that senior pupils in Junior Day Schools of Building will find the study of the book profitable, and that even those craftsmen who have long since passed the stage of apprenticeship will find something to interest them.

As in most craft subjects, good, practical diagrams take an important place in the teaching of Brickwork. Every topic discussed in the book has been clearly illustrated by diagrams which have been specially drawn, and these alone should form a complete pictorial guide for the young craftsman.

Those readers working for examinations will find the book fully covers the City and Guilds of London Intermediate Syllabus in Brickwork.

I should like to acknowledge the help I have received from my colleague, Mr. R. M. Edwards, who has given me the benefit of his friendly criticism, while my thanks are also due to Mr. J. W. Andrews, ACIS, for his help in getting the work into print.

J. C. H.

Contents

1
Craft training

Throughout the 1970s and 1980s there was great pressure for change in the way a craft skill is learned. Brickwork along with other construction industry trades has had its centuries-old tradition of Apprenticeship thoroughly examined.

There are three factors which have caused this re-examination of the Apprenticeship system, with a capital 'A':

(i)   a desire for retraining people who may wish to leave one industry and enter another.
(ii)   a growing shortage of school leavers available and wanting to join construction trades throughout the 1980s, due to a falling birth rate sixteen or so years earlier.
(iii)   a general feeling that just because you 'missed the boat' for vocational training when leaving school, you should not be denied the chance of learning a vocational occupation at any time in later life.

All this is far removed from the traditional arrangement of a school leaver joining a building company for a straightforward period of three or four years' apprenticeship, with attendance at a local College of Technology, as the only way into the construction industry.

Very many patterns of vocational training have been proposed and tried, and continue to be developed at the present time.

Current ET schemes (Employment Training), in operation at College and Training Centres for adult learners, are the Government's response to change-factors (i) and (iii).

Craft skills such as bricklaying, carpentry and plastering, which involve the use of tools and materials and require judgement of hand and eye, should not be confused with assembly processes. The knowledge and practice required to understand how to assemble metal partitions or false ceilings are far less demanding than the skills of setting, cutting, shaping and finishing the materials of the bricklayer, carpenter and plasterer.

Building and Construction's 'Lead-Industry Body', responsible for developing change in training methods, has since 1964 been the CITB. This Training Board appreciates the difference between learning a craft skill in the one case, and that of becoming proficient at an assembly process in the second, by having separate Development Committees for the latter 'Specialist Subcontractors'.

Many operatives in the construction industry enjoyed a traditional Apprenticeship with a caring employer, where a training officer monitored site experience and also progress with further education at a college. Others not so fortunate rather 'endured' their Apprenticeship, which has given the expression 'time serving' a somewhat hollow ring.

The long standing two-part C&GLI (City and Guilds of London Institute) Examinations of Craft and Advanced Craft Certificates based upon syllabuses of practical work and related technology are to be phased out. Objective measurement of a student/trainee's ability in the basic practical competences, step by step assessment of work modules, throughout the weeks of a course of training aims to give a clearer indication of progress to trainer and trainee alike.

Schemes are currently under development: objective assessment of basic skills leading to NCVQ-approved (National Council for Vocational Qualifications), competence-based qualifications, which comprise a prescribed number of units of credit towards the award of an NVQ (National Vocational Qualification).

These, in association with C&GLI and CITB as the Awarding Body, are intended to fit in with the concepts of EC (European Community), with whose policies of training for industry the UK is pledged to integrate.

Whatever form learning a craft skill takes however, it remains an apprenticeship with a small 'a'. Learning a craft remains a developmental process and must still provide sufficient time for repetition in practising the necessary manipulative skills of hand and eye on and off site, together with a sound knowledge of the technology of the trade, and the ability to draw if site plans are to be interpreted.

The student/trainee must realise that formal achievement of basic competences once only, in training, does not indicate total understanding.

Care must be taken by course organisers to see that sufficient job knowledge technology is retained in Units of study leading to NVQs. Student/trainees need not only demonstrate how to carry out a craft operation but understand why it is constructed that way, if they are to gain the in-depth knowledge and ability to satisfy the demands of modern construction processes.

Despite all the changes in the manner and processes of learning a skill, the current uncertainties associated with training and the integration of C&GLI courses within the emerging structure of NVQs, brickwork remains an interesting, satisfying and challenging subject for a career.

2

Materials

Bricks

The study of bricks, from raw materials to delivery of finished products, is an extensive one. At this stage the apprentice should limit understanding to the essentials required in everyday work.

Being able to recognise a brick when it appears on site, know of its properties such as shape, size, weight, strength, porosity, colour etc.—and therefore know how and where to use it correctly—is all-important basic knowledge. Brick making is a very skilful business, with many individual variations in methods of manufacture between companies and their factories.

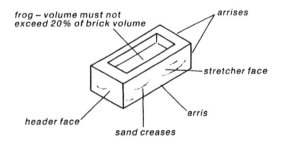

(a) Scale 1 : 10

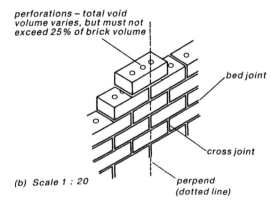

(b) Scale 1 : 20

Figure 2.1 Definitions

Types

Whether made from clay, sand and lime or concrete, bricks may be divided into four broad groups: facings, commons, engineering bricks and refractories.

Facing bricks

These are made in a wide variety of colours and surface textures so as to be durable and attractive to look at.

Commons

These are for general purpose walling which is most likely to be below ground level, externally rendered or internally plastered. They are not given particularly attractive surface features, but are hard and durable. Common bricks have been largely displaced by lightweight building blocks for internal partition walls and the inner leaves of cavity walling.

Engineering bricks

These are exceptionally hard, dense bricks which have a low porosity and therefore absorb very little water. Engineering bricks are intended for walls that are heavily loaded, or very exposed to risk of frost damage. They were originally developed by brickmakers in Victorian times, in response to requests by civil engineers for a very strong brick for use in tunnels, bridges and viaducts. (See 'Compressive strength' below).

Refractories

These bricks are made from specially selected clays which will withstand very high temperatures.

Clay bricks

The British Isles contain extensive deposits of clay and shale suitable for brickmaking. Over one thousand different bricks are produced by UK manufacturers, offering a very wide range of colours and surface textures to choose from. For economic reasons, factories producing clay bricks are usually sited close to or on large deposits of clay or shale.

The natural condition of the raw material determines:

(i)   the type of machinery needed to excavate it;
(ii)   the kind of machine necessary to prepare and temper the clay before shaping it into bricks;
(iii) the need for drying the shaped bricks before firing; and
(iv)   the method of firing to be used.

Stages in clay brick manufacture

No matter how the details of clay bricks vary, and whatever the scale of manufacture—whether in a small works producing ten thousand bricks per week or a mass production company making ten million per week—Fig. 2.2 indicates the five stages which will take place.

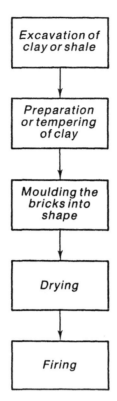

Figure 2.2 Stages in clay brick manufacture

Moulding

Prepared clay can be shaped into 'green' bricks in a variety of ways. The method or machinery used for shaping gives the fired bricks distinctive characteristics.

Pressed bricks are machine made by forcing suitably prepared clay or shale into steel moulds by 'semi-dry' or 'stiff-plastic' processes. Bricks produced by these methods are regular in size and shape, have clean sharp arrises and are smooth, on stretcher and header faces unless gritted or textured before firing.

The semi-dry or fletton process bricks can be recognised by their granular appearance when cut. The stiff-plastic pressing process is used for shaping the clay for engineering brick manufacture.

Solid wirecut bricks are machine made by squeezing and extruding a column of clay through a steel die approximately 225mm by 112mm. This continuous column of clay is sliced into brick thickness by wire frames. The green bricks are frogless and have scratches on the bedding faces due to the cutting action of the wires.

'Dragwire' or 'rustic' facings are produced by scraping one stretcher and both header faces of this clay column immediately it is extruded, and before being wirecut into separate bricks.

Perforated wirecut bricks manufacture is the same as for solid wirecuts, except that a steel comb set inside the mouth of the extrusion machine leaves a continuous pattern of holes through the clay column immediately before it is wirecut into separate green bricks. These perforations permit a more even drying process through the body of the brick and more complete and efficient firing to 1000–1100°C.

Hand made facings Individual bricks of this type are formed by hand throwing a 'clot' of soft clay into a single mould which has been previously sanded so as to release the sticky clay. After cutting off surplus clay by hand, the mould is immediately lifted and the green brick carefully passed on for drying, resting upon its timber base. It is this throwing action which creates the sand creases and soft arrises typical of this desirable and expensive type of facing brick.

Simulated hand made bricks Ingenious machinery copies (simulates) the action of throwing clots of soft clay into six or eight sanded moulds at once. This mechanical throwing action copies that of true hand moulding, making it very hard to tell the difference in the finished fired bricks.

Slop moulded bricks A manufacturing method used by some brick makers whereby the brick moulds are wetted rather than sanded to permit easy and immediate release of the shaped bricks.

Sand moulding is the process described under hand made facings, whereby individual moulds are sanded each time in order to release the shaped green bricks.

Drying clay bricks

After moulding, clay bricks must be dried slowly so that they shrink evenly and without cracking. This process is carried out at modern factories in drying chambers, using recycled heat from the kiln circulating between the green bricks spaced in single layers upon drying racks. See Fig. 2.3.

Some small brickmakers still carry out traditional drying in the open air, with the green bricks stacked under long, low, covered racks called 'hacks'.

Methods of firing clay bricks

Apart from the pure white china-clay deposits, found only in Cornwall, most seams of clay and shale contain various impurities gathered over millions of years of geological time. Iron oxides are impurities which cause bricks to turn red or black during the firing process. Varying the kiln temperature, or the application of sand or other material to header and stretcher faces before firing, can produce single-colour brick types or result in multicolour facings.

Figure 2.3 Typical amount of drying shrinkage of sand moulded clay bricks, from 'green' stage, through dried, to finished fired state

Clamp burning

This is a centuries old method of firing or burning stock bricks. There is no actual kiln structure. Coal dust fuel, approximately 5%, is mixed in with the clay during the tempering stage and before the bricks are moulded into shape.

The clamp is simply a solid stack of dried clay bricks built up upon a 400mm thick layer of coke breeze fuel. There can be a million or more bricks in the clamp, which may have a simple corrugated iron roof structure over it to prevent rain from affecting the burning process.

The base layer of fuel is ignited via 'fire holes' at one end of the clamp, which slowly burns through over a period of weeks. The small percentage of coal dust within each brick contributes to the progression of the firing zone through the clamp.

Some manufacturers supplement the burning process with gas jets inserted along the clamp sides.

After burning through, the clamp of bricks is dismantled and the bricks sorted by eye into first and second quality, on the basis of hardness, colour and shape. Under-fired bricks are used for covering and insulating the next clamp to be built.

Intermittent kilns

These are single chamber kiln structures of brick with walls approximately one metre thick, to reduce heat loss, loaded with dried green bricks; the entrance is temporarily bricked up with a wall built in lime mortar.

The temperature during the firing period of about fourteen days is raised slowly at first to a maximum of approximately 1100°C and held for 7–10 hours. Gradual cooling over a period of days prevents cracks appearing as a result of rapid temperature change, before bricks are removed for inspection.

Hoffman kiln

This is really a terrace block of twenty or more intermittent kilns, all sharing one large chimney stack. Each separate chamber is loaded with about 20 000 green bricks and sealed, before hot air is admitted through flues in the party walls between chambers.

When a chamber has been pre-heated, the firing zone is concentrated there for 7–10 hours, with coal dust fuel added through small fire-holes in the flat roof. This type of kiln is used for fletton brick manufacture and permits a continuous cycle of operations for loading; pre-heating; firing; cooling; and unloading to take place around the total perimeter of the kiln.

Tunnel kilns

With all other types of kiln, the bricks remain in one place during the stages of drying, firing and cooling. With tunnel kilns however, the bricks move through the stages of firing on trollies that run along steel rails laid through the length of the straight tunnel.

The maximum temperature firing zone, approximately half way along the tunnel, is fuelled by gas or oil fired jets operating from the flat roof of these kilns. Each 'kiln car' carries 2000 or more bricks and they are pushed slowly in a continuous train through the length of the tunnel. Kiln cars emerge after a total of thirty hours in the tunnel, to be steadily replaced by others loaded with more green bricks at the opposite end.

Tunnel kilns are installed in most new brick factories because of their accurate, computer controlled firing temperatures and general efficiency.

Classifying bricks

BS3921:1985 (Specification for Clay Bricks) provides manufacturers and architects with a more scientific classification of clay bricks and their properties than the descriptions given in Table 2.1.

Table 2.1   Some examples of traditional classification for clay bricks

Brick description	Classified by
Staffordshire blue engineering brick	Location Colour Use
Handmade red facing	Method of manufacture Colour Use
Solid wirecut common	Method of manufacture Use
Sand faced fletton	Surface texture Method of manufacture
Perforated wirecut dragwire buff facing	Method of manufacture Surface texture Colour and use
London stock	Location Method of manufacture
Handmade sandfaced multi-colour Dorking stock	Method of manufacture Surface texture Colour Location

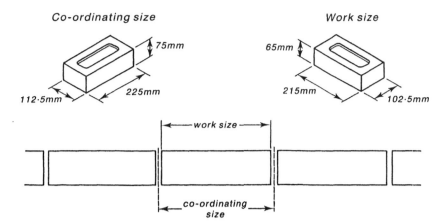

Figure 2.4 Sizes of standard metric bricks: co-ordinating (nominal) and work (actual)

Size of standard metric bricks

All clay brick manufacturers aim to produce their standard metric size bricks to these work size dimensions. Clay is a naturally occurring material, and variations in drying shrinkage mean that actual dimensions can vary from the intended work size; see Fig. 2.4.

A mortar joint all round each brick of approximately 10mm allows bricklayers to stick rigidly to the co-ordinating size of 225mm × 112.5mm × 75mm, by adjusting the mortar joint thickness as necessary.

Dimensional deviations

BS3921 imposes limits on dimensional variations of clay bricks with a 24 brick test as illustrated in Fig. 2.5, using bricks taken at random from a delivery.

Compressive strength

Bricks may be specified by compressive strength depending upon where they are to be used in a building. Clay bricks can vary from $5N/mm^2$(Newtons per square millimetre) up to $100N/mm^2$ compressive strength.

Engineering bricks which have a compressive strength of not less than $50N/mm^2$ are called Class B, and may be either solid or perforated types. Those of more than $70N/mm^2$ are called Class A engineering bricks and are usually solid types. Average-strength facing bricks will have a compressive strength of about $20N/mm^2$. BS3921 requires the crushing strengths of ten sample bricks to be averaged when testing for compressive strength.

Water absorption

Very dense engineering bricks may be used for a damp proof course (dpc), instead of flexible bitumen materials. Those bricks with only 4.5% by weight water absorption are suitable as a dpc in buildings. Those having

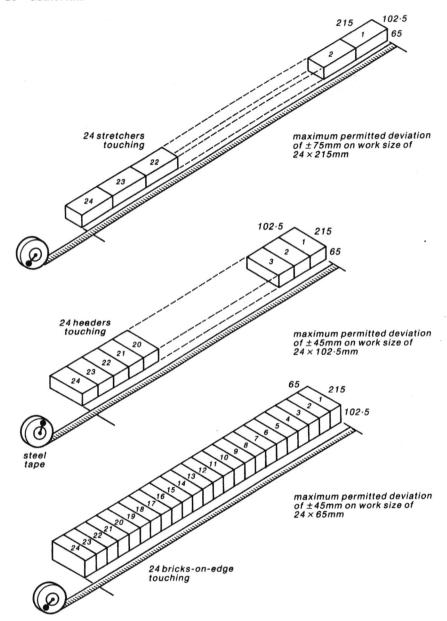

Figure 2.5 Carrying out a BS3921 test for dimensional deviations

not more than 7% absorption are adequate for a dpc in garden walls and other external works. BS3921 requires the average of ten sample bricks to be tested when checking for water absorbancy. See Table 2.2.

Table 2.2 BS 3921 classification of bricks

Class	Compressive strength in N/mm^2	Water absorption % by mass
Engineering A	Not less than 70	Not more than 4.5
Engineering B	Not less than 50	Not more than 7
DPC type 1	Not less than 5	Not more than 4.5
DPC type 2	Not less than 5	Not more than 7
All other bricks	Not less than 5	No limits are set

Frost resistance of bricks

Bricks are classified into degrees of resistance to frost, which is printed on the packaging of bricks as delivered to sites or otherwise indicated.

'F' Frost resistant designated bricks are durable in all building situations and can withstand freezing even when the wall is saturated.

'M' grade Moderately frost resistant bricks are durable except when in a saturated condition and exposed to freezing.

Bricks that do not meet with 'F' or 'M' grades are classed as 'O'. These are not resistant to frost and need covering during construction and permanent protection from the weather, eg they are suitable for internal walls or behind cladding in a finished building. Manufacturers do not set out to make 'O' grade bricks.

Soluble salt content in clay bricks

Other impurities in clay raw material from which bricks are made include various soluble salts. These salts are impossible to remove before bricks are made, and can cause problems of efflorescence and sulphate attack on cement mortar in walls which may be wet for long periods each year in completed buildings.

It is important that bricks with a low soluble salt content, 'L' grade, are specified for walls below ground level, retaining walls, parapets and chimney stacks.

Bricks given an 'N' grade rating, with normal soluble salt content, could be at risk only if used in walling exposed to continual dampness.

Durability of clay bricks

The two properties of frost resistance and soluble salt content are brought together in BS3921, to give the architect/specifier an indication of the long-term durability of bricks, i.e. an 'FL' designation would indicate that the brick is resistant to frost and low in soluble salt content. See Table 2.3.

Table 2.3 Durability (BS 3921)

Brick designation	Frost resistance	Soluble salt content
FL	Frost resistant (F)	Low (L)
FN	Frost resistant (F)	Normal (N)
ML	Moderately frost resistant (M)	Low (L)
MN	Moderately frost resistant (M)	Normal (N)
OL	Not frost resistant (O)	Low (L)
ON	Not frost resistant (O)	Normal (N)

Table 2.4   Calcium silicate bricks — compressive strength classes

Class	Mean compressive strength not less than (N/mm^2)	Colour marking of packs
7	48.5	Green
6	41.5	Blue
5	34.5	Yellow
4	27.5	Red
3	20.5	Black

Note: For comparisons, 1000 psi $\equiv$ 7N/mm^2

Calcium silicate bricks

More commonly called sand lime bricks on site, these are made from a carefully controlled mixture of 90% fine silica sand plus 10% lime. This mixture is pressed into steel moulds and then steam hardened.

Coarse flint sand is also used to make flint-lime bricks, which after steam curing have a higher compressive strength, see Table 2.4.

Loaded trollies are pushed into steel steam chambers called autoclaves which hold a total of 22 000 bricks. Steam pressure is maintained for seven hours to cause the sand and lime to fuse together. Upon cooling sample crushing tests allow the whole chamber to be given a compressive strength grading which may be between 21N/mm^2 and 49N/mm^2.

Colours

The basic colour of calcium silicate bricks, which are very smooth and of regular shape, is white. Different powder pigments added during the mixing stage of manufacture gives a range of facing brick colours. Multi-colour and rustic surface texture calcium silicate bricks are also available.

Size

Calcium silicate bricks are manufactured to the same work size as clay bricks, namely 215 × 102.5 × 65mm, and may be solid or frogged. They contain no soluble salts and are frost resistant.

Compressive strength

Identical looking white calcium silicate bricks of different compressive strength batches are identified by paint colour-marking of the packs as indicated in Table 2.4. The strength class of calcium silicate bricks, shown by number in Table 2.4, indicates pre-metric thousands of pounds per square inch, (i.e. Class 4 indicated 4000 psi etc). (*Note* 1000 psi is approximately equal to 7N/mm^2).

Concrete bricks

These are cast from a mixture of fine aggregate and Portland cement pressed into steel moulds. The natural setting and hardening process of cement determines the compressive strength of these bricks. They may be solid or frogged and are available as facings, commons and engineering quality.

Colour

The basic colour of concrete bricks is grey due to the Portland cement. Inorganic powder pigments are used to produce a range of plain and multi-colour bricks.

Surface texture

Concrete bricks are made smooth faced, exposed aggregate weathered, or split so as to resemble natural stone.

Size

The standard metric size is practically the same as for clay bricks, being $215 \times 103 \times 65$mm. Metric modular bricks are 190mm and 290mm long to give courses 65mm and 90mm deep and are available to order.

Compressive strength

Facings and commons are made with a compressive strength of $21N/mm^2$; engineering quality at $40N/mm^2$.

Refractories

More commonly referred to as firebricks, these can be considered as two sorts:

(i) very dense cream coloured solid bricks used to contain the fire in furnace linings, cement kilns, ships' boilers and for lining steel ladles transporting molten metal in a steel works; and
(ii) very lightweight bricks used to 'back-up' dense refractories, or to insulate chimney shafts in order to prevent heat escaping from flue gases. (This is the only brick that will float in water!)

Manufacture

Selected naturally occurring deposits of fire clay are moulded, dried and fired like other clay bricks to produce dense refractories able to withstand temperatures between 900 and 1350°C. Other fireclays containing a higher percentage of aluminium produce dense refractories able to withstand temperatures up to 2050°C.

The second sort of very lightweight refractories used for heat insulating purposes around the lower section of large chimneys, are made from a special fireclay called diatomaceous earth.

Size

A standard size dense firebrick of $230 \times 114 \times 76$mm is accompanied by a range of arch shapes, wedges and bullnose bricks made to suit circular section kilns and boilers. Cellular blocks of this same dense refractory material are used to protect the steel deck of the brick-transporter cars used in tunnel kilns.

General terms

Flettons

The word 'fletton' is no longer a registered trade mark. It is used to describe bricks made from a hard, dry shale which is finely ground and forced into steel moulds under great pressure. This semi-dry process of manufacture makes use of a wide range of sanded surface textures applied to header and stretcher faces. Flettons are the only clay bricks made from raw material that contains a small percentage of naturally occurring shale oil, which assists and reduces the cost of the firing process of these bricks.

Stock bricks

This is a term that means different things to different people. It can be guaranteed to cause arguments between bricklayers! By tradition, stocks are manufactured in SE England, using clays prepared to a very soft consistency. Originally they were hand thrown into a 'mould box' set over a timber 'stock', which formed the frog of each brick.

A small percentage of coal dust or coke breeze is continually blended into the clay as it passes under the heavy rollers of the pug mill and before the moulding stage. This small addition of fuel assists the firing stage of manufacture, which again by tradition has been carried out by clamp burning.

This method of firing, being difficult to control scientifically, means that stock bricks have always needed visual sorting and grading into first and second quality—for shape, hardness and colour consistency. Rejected bricks are often used to cover and insulate the next clamp to be built.

Millions upon millions of yellow 'London Stocks' were produced for Victorian buildings in south-east England, and they continue to be made using modern moulding and firing processes. Having a similar method of manufacture to 'Yellow London Stocks', 'Kentish Stocks' have a red-brown multi-coloured appearance, and an equally long history.

Mortar _____

It is not usually difficult to cause an argument between bricklayers; ask them the following question. Does mortar stick bricks together or keep them apart? The answer is of course that it does both things, but a lot more besides. Mortar must stick firmly to bricks and blocks in external walling so as to keep the rain out. Mortar bed joints also hold bricks apart, so that the courses can be kept level and to an even vertical gauge of four courses to 300mm, with standard metric bricks.

When walls were much thicker and Portland cement had not been invented, a mortar mix of lime and sand was used very successfully for thousands of years. The lime used then was hydraulic; meaning that it was made from a naturally occurring chalk raw material which contains clay impurities. This clay content gave the lime a slow setting action not unlike the chemical hydration-setting-hardening process of ordinary Portland cement, but, the lime mortar would take years to harden.

Very pure or 'fat' limes are used in plastering and other industrial processes and also by gardeners.

Vitruvius, writing a textbook in Roman times, gave careful instructions for preparing lime mortar. Ask your nearest librarian to obtain a copy of an English translation for you.

Raw materials

Sand

The bulk or mass of mortar is a fine grain, clean sand, usually described by builders' merchants as 'building sand'. Many grains of fine sand have a greater total surface area for retaining mixing water than the same volume of coarser grains. Fine sand gives a neater smoother joint finish as well.

All sands should be supplied 'graded'. This means that they must contain a range of particle sizes from large to small, so as to give the hardened mortar maximum density. All sands for use in construction should be washed so as to remove mud or silt. Muddy, unwashed sand results in a weak mortar, because this prevents the cement sticking firmly to each grain of sand. Coarse grain, graded, sharp sand, with a higher percentage of particles closer to the 5mm maximum size, will produce an unworkable bricklaying mortar, but may be ideal for use as a fine aggregate in concrete.

Cement

If you read the writing on a cement bag you will see that its correct name is Ordinary Portland Cement. This is the most commonly used cement on construction sites for making mortar and concrete. It is the all-important 'glue' in mortar which binds the grains of sand together when water is added and the setting and hardening process is completed. Used neat on its own, cement is too sticky, sets too hard and would develop severe shrinkage cracks. It is therefore always diluted with three, four or six equal volumes of sand.

The setting and early part of the hardening processes of Portland cement involve complicated chemical reactions between the mixing water and the cement powder in a batch of mortar or concrete. These reactions need to take place in damp conditions, called 'curing', if they are to be totally completed and give full hardened strength.

This early setting and hardening (see Fig. 2.6), taking place over hours and weeks respectively, must never be hurried by early 'drying out', as this will severely reduce the final strength of ordinary Portland cement mortar or concrete.

Figure 2.6 shows the stages in hydration of OPC as it sets and hardens, whether used in mortar or concrete.

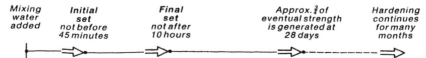

Mixing water added | Initial set not before 45 minutes | Final set not after 10 hours | Approx. ⅔ of eventual strength is generated at 28 days | Hardening continues for many months

Figure 2.6   Typical setting and hardening timescale for ordinary Portland cement

Plasticizer

Generally speaking bricklaying mortar made from cement and sand only is
not sufficiently fatty or easily workable with the trowel. Such mortar is
described as 'short' or 'harsh' and does not hold together when rolled on
the spot board (see Fig. 7.1). Lime added to a mix improves the workability
by temporarily retaining more mixing water, and results in a denser mortar.
Lime is usually added in a volume equal to that of the cement in gauged
mortar, i.e. 1:1:6. As an alternative, patent liquid plasticizers can be added,
which generate millions of micro bubbles of air within the mortar as it turns
in the mixer. Liquid plasticizers must be added by the mixer driver on site
with great care—following the manufacturer's instructions, printed on the
container. These are called air-entrained mortars.

Water

Mixing water triggers off a chemical reaction with cement, which causes the
setting and hardening of mortar and concrete. Water should be clean
enough to drink, as impurities can seriously delay or prevent this setting
action.

Purpose

Bricklaying mortar is the ideal material for getting bricks to rest firmly
upon each other, whether these are accurately shaped Class A clay
engineering and calcium silicate bricks, or more irregularly shaped hand
made bricks.

The mortar must remain soft enough for each brick to be pressed down
to the line, before suction causes the bed to stiffen up. Not only does
mortar accommodate irregularities, but it must stick firmly to each brick so
as to stop rain from penetrating exposed joints.

Types of mortar

The basic raw materials for bricklaying mortar can be prepared in a
number of different ways, depending upon specification and site requirements. The apprentice/trainee should understand the definitions given in
Table 2.5.

Mortar designation groups

BS5628:1985 divides bricklaying mortars into five groups for reference
purposes, using Roman numbers (i) to (v) (See Table 2.6 and Fig. 2.7). The
Bricklayer section of a Job Specification or Bill of Quantities is just as
likely to specify mortar for brick or blocklaying by such a Group
Designation number as it is to state mix proportions such as 1:1:6 or 1:1:5.
Figure 2.8 shows that Designation group (iii) mortar can be produced with
cement/lime/sand, *or* masonry cement/sand *or* as a plasticised mortar, each
having approximately similar compressive strength when hardened.

The stronger mortar, Designation group (ii) can similarly be produced in
any one of these ways, but using less sand in each batch as indicated by the
volume proportions.

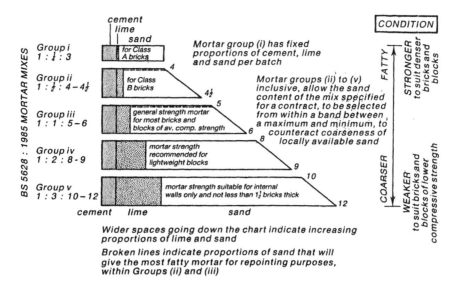

Figure 2.7 Cement/lime/sand mortar—another way of looking at mortar designations for brickwork and blockwork

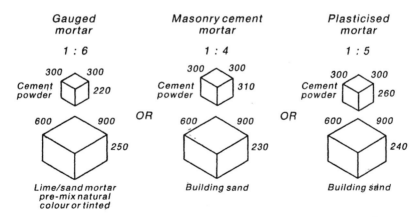

Figure 2.8 Three ways of producing a designation group (iii) mortar

Table 2.5 Types of bricklaying mortar

Mortar type	Definition	Advantages	Disadvantages	Remarks
Lime mortar	Mortar made from 1 volume of lime to 3 equal volumes of sand	Smooth and workable. Ideal for training purposes where materials must be recycled	Unsuited to modern construction particularly cavity work; due to slow rate of hardening and final strength	1:3 mix using hydraulic lime, recommended by conservationists for repairs and repointing centuries old brickwork
Cement mortar	Mortar made from 1 volume of OPC to 3 equal volumes of sand	Recommended only for use with Class A engineering bricks	Stiffens and sets too rapidly. Will cause shrinkage crack damage to other bricks and blocks if used by mistake	Usually has $\frac{1}{4}$ volume of lime added to improve workability
Compo	Composition mortar using separate site deliveries of dry OPC, lime powder in bags, and sand in bulk	Combines quicker strength gain of OPC with good workability of lime	Messy splitting of bags of cement and lime powder on site, giving health and safety risks	Obsolete for site use due to disadvantages given
Gauged mortar	Mortar prepared on site from bulk deliveries of wet-mixed LSM, to which 1 equal volume of OPC is added or 'gauged' each time a batch of mortar is required	Added powder pigments can be carefully measured and blended at works, before delivery to site, to ensure consistent colour of hardened mortar	Bulk delivery or stock-pile of LSM, to last a week or so, must be kept covered to prevent surface drying in summer and pigment blowing away. This, and rain washing, can weaken the colour strength of mortar batches	A reliable way of producing tinted bricklaying mortar

Masonry cement mortar	A purpose made cement for use only in bricklaying mortar. Mixed on site with 5 equal volumes of sand for general brickwork above dpc level	50kg bags contain powder plasticiser ready mixed in with OPC powder. Does *not* require addition of lime or liquid plasticiser to batches of mortar	Masonry cement must *not* be used to make concrete	Unsuitable for concrete because it has been diluted with powder filler
Plasticised mortar	A cement/sand mortar to which a purpose-made liquid plasticiser is added at the site mixer to improve workability	A more convenient way to improve workability; easier to store and use than bagged lime	Excess mixing time will over-generate micro bubbles and result in a sloppy batch of mortar	Micro bubbles lubricate grains of cement and sand. Washing-up liquids must *not* be used instead of purpose-made plasticiser, as these can affect durability of mortar
Ready to use mortar	A complete RTU mortar delivered to site pre-mixed in plastic tubs. This CLM contains a chemical retarder which delays the setting action for 36 hours whilst in the tubs	No site mixer required. Weigh batching at suppliers depot removes problem of colour and strength variations between batches, which can arise with site mixed mortar	Difficulty of predicting exact daily requirements of RTU mortar, so as to avoid waste or bricklayer waiting time. Supply depots normally need a day's notice for deliveries	Full range of tinted RTU mortars available from suppliers in addition to natural colour

(*Abbreviations*: CLM—cement lime mortar, LSM—lime sand mortar, OPC—ordinary Portland cement, RTU—ready to use)

Table 2.6 Mortar designations with average compressive strengths

| BS 5628 Pt 3 Mortar designation number | Type of mortar | | | Average compressive strength in N/mm² at 28 days site testing |
| | Cement/lime/sand composition mortar | Air entrained mortars | | |
		Masonry cement/sand mortar	Plasticised cement/sand mortar	
(i)	1 : 0 to ¼ : 3			11
(ii)	1 : ½ : 4 or 4½	1 : 2 or 3	1 : 3 or 4	4.5
(iii)	1 : 1 : 5 or 6	1 : 4 or 5	1 : 5 or 6	2.5
(iv)	1 : 2 : 8 or 9	1 : 5½ or 6½	1 : 7 or 8	1.0
(v)	1 : 3 : 10 or 12	1 : 6.5 or 7	1 : 8	

Note: Numbers shown in the three middle columns, indicate parts measured by volume for the different mortar mixes. The symbol N/mm² indicates Newtons per square millimetre stress.

Selection of mortar

As a general guide, the hardness or eventual compressive strength of mortar should be related to the hardness of, or preferably slightly weaker than, the bricks or blocks to be laid. So that if, as a result of slight foundation settlement, cracks develop, these will follow the joint lines, which can easily be cut out and re-pointed. Excessively hard mortar can result in the bricks becoming fractured at settlement cracks, thereby leading to a more extensive repair operation. Another reason for relating strength of mortar to the compressive strength of the bricks, is to ensure that external walls weather evenly during the lifetime of the building, with any absorbed water evaporating at a similar rate from the surfaces of bricks and joints alike.

In addition to being hard enough to transfer loads evenly between irregular surfaces of bricks, the choice of mortar must resist the effects of rain and frost in the long term, see Table 2.7.

Mixing mortar

Machine mixing is the most effective way of turning cement/lime and sand dry materials into mortar on site. A typical small tilting drum mixer is shown in Fig. 2.8. The mixer driver must be instructed in the importance of using gauge boxes of the correct size for every batch of mortar produced, and given the reasons why it is important to do so.

Table 2.7 Mortar requirements

Good workability	Smooth and easy to handle with the trowel when transferring from spot board to wall, and applying cross-joints
Water retention	Mortar should contain sufficient fine particles of sand, together with cement and lime to prevent mixing water 'bleeding out' on the spot board
Adhesion	Mortar must stick to bricks and blocks to prevent rain penetration of joints
Durability	Mortar for externally exposed brickwork must resist the combined effects of rain, frost and any soluble sulphate salts in bricks of fired clay. (See Table 2.3)

The description of this standard site mixer as a '5/3½' indicates the volume of dry materials put in and the volume of wet mortar discharged per batch (see Fig. 2.8).

Mixing water

The water added to each batch of mortar is not measured in the same way that an exact water/cement ratio controls what is added to a batch of concrete. A typical water/concrete ratio for concrete is 0.5. This means that the amount of mixing water allowed is 50 per cent of the weight of cement powder per batch. For example, with 50kg cement, only 25kg of water is permitted (25 litres). Too much water weakens concrete.

Mixing mortar is a matter of what feels right: it must be workable with the trowel: too dry and it will not spread into bed joints properly, too wet and it smears the face of the bricks. The experienced mixer driver knows the workability or consistency that bricklayers require, and he will soon be told if it is coming out wrong!

Batching

The proportions of dry materials for each mix of mortar must be carefully and regularly measured/batched if the strength and colour of the hardened mortar is to be consistent. Variations can seriously affect durability of brickwork and result in conspicuous patchiness on completed facework elevations.

Bricks and mortar walls near the coast for example will need to be more resistant to the eroding effects of the weather than those sheltered by other buildings in towns and cities. Colour-matching or contrasting joints with the facing bricks, from the range of tinted mortar colours available, is another consideration.

Bulk deliveries of tinted lime-sand mortar, to which cement is added in carefully regulated proportions using a gauge box, to produce 'gauged mortar', or daily deliveries of tinted 'ready-to-use' mortar, are both recommended methods.

Large scale use of powder pigments added to the site mixer is not a reliable method, due to the difficulty of ensuring colour consistency of successive batches.

Measuring dry materials by the shovelful is totally unsatisfactory, as each will hold a different volume of cement powder or sand. Figure 2.9 shows typical bottomless gauge boxes, the regular use of which will ensure that measurement or batching remains consistent where mortar is mixed on site.

Placed on a flat surface behind the mixer, the larger is filled with sand and struck off level. The smaller, placed on the flat sand surface, is filled level with cement powder.

Both gauge boxes are lifted and the total contents transferred to the mixer by shovel. Fig. 2.8 indicates the size of gauge boxes that will produce a Designation group (iii) mortar which will fill the standard size mixer used on sites.

Figure 2.9   Typical gauge boxes

Retarded ready to use mortar

Photograph Fig. 2.10 shows a typical delivery of retarded ready-to-use bricklaying/blocklaying mortar, direct to sites.

The cement, lime and sand dry materials are mixed at a factory depot, complete with the required amount of water for optimum workability, plus a chemical retarder. This delays the onset of the initial setting action for 36 hours, whilst the mortar is in the delivery tubs. Upon delivery, these $0.3m^3$ (600kg) plastic tubs of mortar can be transported by fork-lift truck or tower crane directly to the point of use. Ready-to-use mortar avoids the risk of possible carelessness with site batching that can be the cause of colour and strength variations in finished brickwork. Bulk deliveries of lime/sand mortar plus bagged cement and a mechanical site mixer are all unnecessary if ready-to-use mortar is specified, a useful consideration on congested city centre sites.

Mortar testing

Increasingly, quality control procedures call for compressive strength tests to be carried out on a regular basis, throughout the bricklaying operations

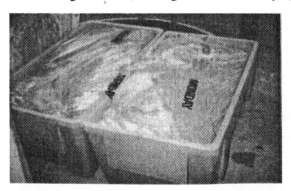

Figure 2.10   Delivery of retarded cement/lime/sand premix mortar. Note the delivery day marked on each polythene cover, to ensure use in rotation

Table 2.8 General recommendations for mortar mixes

Use	BS 5628 Mortar designation number	Cement/ lime/ sand mortar	Types of mortar Masonry cement/ sand mortar/	Plasticised cement/ sand mortar
		(parts by volume)		
Class A and B engineering quality brickwork	(i)	1:0 to $\frac{1}{4}$:3	—	—
Work up to dpc level, including extremely exposed brickwork eg chimney stacks, parapets, free-standing boundary walls, BOE sills and copings	(ii)	1:$\frac{1}{2}$:4	1:3	1:4
External facing and common brickwork above dpc level and internal walls of medium density blockwork	(iii)	1:1:6	1:4	1:5
Internal walls of lightweight blockwork	(iv)	1:2:8	1:6	1:7

on site. The same steel moulds familiar for testing concrete, are used to make 100mm × 100mm × 100mm mortar cubes for laboratory testing at 7 day and 28 day intervals. The Bricklayer section of a Job Specification will describe these mortar test requirements, and whether a copy of the mortar suppliers' own test results will be acceptable, or if an independent laboratory is to be appointed.

Typical compressive strength of different mortar mixes are given in Table 2.6. These are results based upon site tests, and will show a continuing slight increase in the weeks following a 28 day test.

It is worth noting that typical compressive strength of ready-to-use mortar is ultimately 25–30% higher, due to better batching control by weight rather than volume, and also because the slowed rate of initial setting and hardening improves hydration and eventual strength gain of the cement in the mortar.

Sulphate attack on Portland cement mortars.

The use of sulphate resisting cement, prevents sulphate attack from developing in any brickwork that may be saturated for long periods of time (see Table 2.8). Sulphate resisting cement will therefore be required for brickwork below dpc level in those subsoils which contain water soluble sulphates.

Chimney stacks, parapets and free-standing boundary walls constructed of fired clay bricks, manufactured from a brick-earth that contains soluble sulphates, will also be at risk unless sulphate resisting cement is specified.

Reinforced mortars

This is a term used to describe those admixtures such as sytrene butadiene, added to cement/lime mortars to improve adhesion and water resisting properties of brick-on-edge sills and copings. Great care must be exercised to follow the manufacturers' instructions carefully if these admixtures are to work effectively and be used safely.

3
Tools

Tool kit

Bricklayers can carry their complete tool kit (Fig. 3.1 to 3.3; see also Fig. 14.2 for jointing and pointing tools) in a medium size canvas shoulder bag or heavy duty hold-all in one hand, and move about site with a spirit level in the other. A kit of tools is very personal property and the apprentice will become familiar with all the items, particularly the brick-laying trowel and will not want other bricklayers to use it.

A tool kit must be maintained in good order so as to ensure efficient and safe working, avoiding accident to self and others. On construction sites safety helmets must be worn at all times for brain protection—we only have one! Eye protection must be worn when cutting brickwork, blocks or concrete to protect irreplaceable eyes—we only have two!! Feet should be protected with stout safety shoes or boots. Soft top track shoes give no protection against a hammer accidentally dropped from hand height, never mind anything worse. Do not throw your tool bag down, as this will damage and loosen handles.

Trowels should be washed clean at the end of a day's work, scouring round the shank with a small piece of brick. Spirit levels should also be washed each night and checked for accuracy every week.

There is no excuse for a bricklayer to be uncertain about the accuracy (or otherwise) of a spirit level. The simple checks for level and plumb, shown in Figs 3.4 and 3.5, can be carried out at any time, should a spirit level/ plumb rule be knocked or dropped.

Reversing a level

Checking the horizontal bubble

If a course of bricks, a sill or a lintel has been set perfectly horizontal, 'reversing the level' end-for-end should confirm this; see Fig. 3.4.

If the spirit bubble reads truly horizontal at Fig. 3.4, but is *not* horizontal when the level is reversed, then the spirit level is out and needs to be adjusted. The clamping screws for the horizontal bubble tube must be slackened, and the necessary adjustment made.

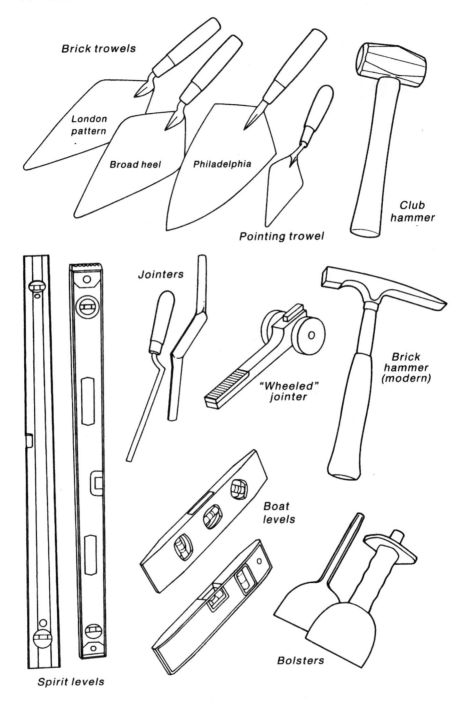

Figure 3.1 Bricklaying tools

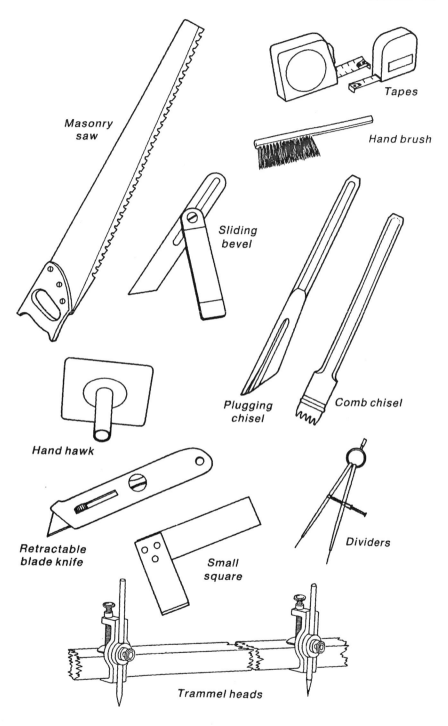

Masonry saw

Tapes

Hand brush

Sliding bevel

Plugging chisel

Comb chisel

Hand hawk

Retractable blade knife

Small square

Dividers

Trammel heads

Figure 3.2 Bricklaying tools, continued

Checking the plumbing bubble

(i)   Place the spirit level upright and flat against a wall surface.

(ii) Slide the level sideways on the wall surface so that it reads exactly plumb.

(iii) Make a small pencil line top and bottom, as Fig. 3.5(a).

(iv) Roll the level over, and slide the same edge back to coincide with the lower pencil mark, as Fig. 3.5(b).

(v)   Move the upper part of the spirit level so that it reads plumb, and again mark the top position in pencil.

(vi) If the two upper pencil marks coincide, then the spirit level is accurate for vertical use.

(vii) If there are *two* pencil marks side by side at the top, see Fig. 3.5(c), then the level is not reading truly plumb. (Any error is half the distance between the top two pencil marks in every metre of plumbing; the error is cumulative the higher the building progresses). The clamping screws for one or both of the bubble tubes used for plumbing must then be slackened and adjustments made.

Hammer heads should be refixed directly they loosen and combs in scutches replaced when worn. Bolster and cold chisels must be regularly sharpened to ensure efficient operation, and early developing 'mushroom heading' ground off frequently to prevent risk of nasty injury from steel splinters.

Bricklayers need to think about security of their tool kit. It should be locked away overnight and brought home if away from site due to illness or holiday.

Brick trowel

This is the most heavily worked item in a bricklayer's tool kit, used for gathering and spreading mortar, applying cross joints and for rough cutting some kinds of brick. Available in a range of shapes, sizes and thickness of steel, with length of blade from 230mm to 330mm. Choice of trowel has a lot to do with personal preference and what feels most comfortable for the individual. The apprentice should avoid the temptation that biggest is best.

Narrow blade 'London pattern' trowels are suited to cavity work, broad heel trowels lend themselves to solid walling.

The best trowels are solid forged from a single piece of steel from tip to tang. The majority have one side of the blade of thicker steel to withstand wear and tear of tapping bricks and rough cutting. For this reason, left and right handed versions are produced.

Philadelphia pattern brick trowels are not intended to be used for rough cutting, and so are not 'handed'. Use of the trowel handle to tap bricks down to the line should be avoided as this tends to loosen the handle and 'burr' the end, making it uncomfortable to use.

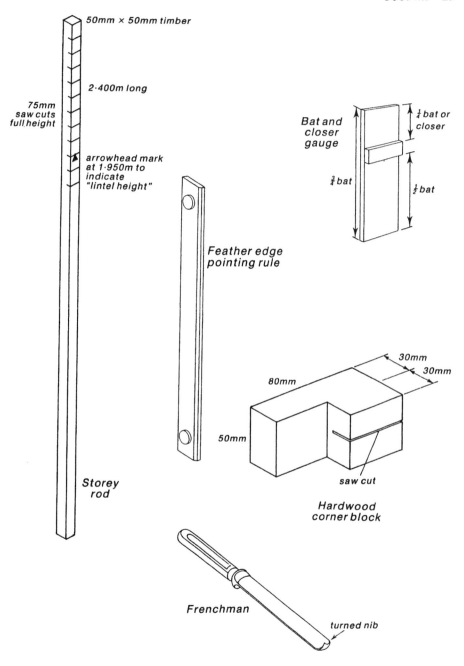

50mm × 50mm timber

2·400m long

75mm
saw cuts
full height

arrowhead mark
at 1·950m to
indicate
"lintel height"

Storey
rod

Bat and
closer
gauge

¼ bat or
closer

¾ bat

½ bat

Feather edge
pointing rule

30mm

30mm

80mm

50mm

saw cut

Hardwood
corner block

Frenchman

turned nib

Figure 3.3 Bricklaying tools, continued

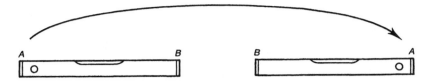

Figure 3.4 Daily spirit-level check: horizontal level

Pointing trowels

These are made with blade lengths from 75mm up to 175mm. That with the shortest blade sometimes referred to as a 'dotter', is used for filling and striking cross joints. The longer blade pointing trowels are used with a hand hawk for filling and striking bed joints.

Spirit levels

These have a dual purpose and are used for checking the horizontal and vertical accuracy of brickwork. For general purpose use, a bricklayer's spirit level should be not less than one metre in length, preferably 1200mm long. Available as girder section hardened aluminum with hand-hold slots, or as a hollow box section enamelled aluminum, or in hardwood.

A professional bricklayer never taps or knocks a spirit level when plumbing or levelling brickwork. 600mm long spirit levels are made for use in restricted places. 200mm long boat levels are made for plumbing soldier bricks and for decorative brickwork.

Line and pins

No tool kit is complete without a 'set of lines' for controlling level, line, plumb and gauge of any walls over 1.200m long. Always buy the best quality hardened steel pins you can afford so that they last.

Bricklayer's line made from traditional hemp can be spliced if accidentally broken so as to avoid knots. Cotton, polyester and various types of nylon are also sold, but it is impossible to join any break in these materials by splicing, as the strands do not separate neatly. The finer the line the better for accurate work. When renewing line, wrap insulating tape around the pins first to prevent rust staining. Always tie the line-end on to pins before winding-on, in case you drop the pin when working on a high scaffold.

Jointing tools

A variety of tools are shown for applying a permanent finish to the exposed surface of mortar joints, some purpose made, some produced by the bricklayer. Patent wheeled jointers are ideal for raking out mortar to a constant depth in preparation for square recessed joint finish.

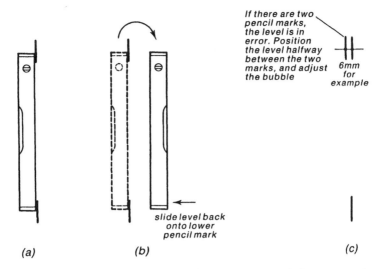

Figure 3.5 Daily spirit-level check for plumb; (a) see text (i) to (iii); (b) see text (iv) and (v); (c) see text (vii)

Club or lump hammer

One kilogram size is ideal for use with a bolster for fair cutting and also for cutting away existing brickwork with a cold chisel. These have ash or hickory-wood handles or of steel with a rubber sleeve.

Bolster

A 100mm width blade is most useful for fair cutting bricks, and bolsters can be supplied with or without rubber or plastic collars for hand protection. Bolsters must be kept sharp for efficient operation.

Brick hammer

Used for rough cutting very hard bricks which would damage a trowel. They have one hammer head and one forged chisel end which can be reground when worn.

Comb hammer

This has a hammer head one side and is slotted to take hardened steel replacement combs at the other side.

Scutch

This is slotted both sides to receive renewable hardened steel combs or blades.

Measuring tapes and rules

Folding boxwood or plastic rules tend to get broken easily and have been largely replaced by steel tapes. These also have a limited life as they tend to get full of grit and do not fully retract.

Cold chisels

These must be kept sharp if cutting away brickwork is to be as painless as possible. Grind-off the first beginnings of burring over, don't wait until a full grown and dangerous 'mushroom' has developed. Always use a steel point and eye protection if you are unfortunate enough to be cutting away concrete, don't ruin a cold chisel.

Comb chisel

A very useful substitute for a cold chisel when cutting away brickwork.

Plugging chisel

A purpose made tool for carefully cutting and toothing out joints in existing brickwork (see Fig. 3.2).

Hawk

Allows mortar to be picked up conveniently with a pointing trowel. The removeable handle makes it easier to fit into a tool bag.

Hand brush

Medium to fine bristle is used for lightly brushing face brickwork at the end of a day's work. Great care must be taken not to leave bristle marks in any mortar still soft.

Small square

For pencil marking bricks accurately before cutting.

Dividers

For spacing out voussoir positions on an arch support centre (see Chapter 8), either side of the key brick location before starting work.

Bricklayers' sliding bevel

Used for transferring and marking the same angle of cut for all the bricks when raking cutting to a gable-end wall or when carrying out tumbling-in.

Masonry hand saw

With tungsten carbide tipped teeth for toothing out existing brick or blockwork where the mortar is not too hard, and also for neatly cutting through lightweight aerated concrete blocks.

Pair of trammel heads

One a steel compass point, the other for holding a pencil. For use with a timber lath as a beam compass, when making a template for constructing a curved wall, drawing arches full size or striking the shape of the brick core to a bullseye.

Pencil

For plumbing perpends on face brickwork, H or 2H grade pencils will last longer than HB grade.

Retractable blade knife

Very useful for cutting dpc material and sharpening pencils.

Items of equipment additional to the tool kit _____

These may be made up by the bricklayer.

Brick cutting gauge

For quick accurate marking of standard size cut bricks required for bonding purposes. This item should be made from oak or other hardwood in order to survive the rigours of life in a tool bag.

Storey rod

Indispensible for building and checking courses of brickwork and block work, so as to ensure consistent vertical gauge above and below datums.

Corner blocks

A simple means of supporting bricklayer's line at quoins that does not leave pinholes behind. If made from hardwood will last longer.

Frenchman

A simple device made from an old table knife, heated, bent over and filed to shape for trimming mortar bed joints when carrying out weather-struck and cut pointing, or tuck pointing.

Feather edge pointing rule

Used in conjunction with a Frenchman for trimming bed joints after pointing. Cork pads allow trimmed mortar to fall away cleanly.

4

Bonding of brickwork

The majority of brickwork produced is in the form of cavity wall construction for the external walling of buildings; (see Chapter 9). Stretcher bond makes the most economic use of expensive facing bricks in the half-brick thick (102.5mm) outer leaf of this type of construction, see Fig. 4.1.

It is, however, most important for the apprentice to have a wider understanding of the basic principles and rules of bonding which are best illustrated on *solid* walling greater than 102.5mm thick. The apprentice will encounter non-cavity or solid brickwork in freestanding boundary walls, building refurbishment, earth retaining walls, chimney breasts, civil engineering brickwork, inspection chambers, etc.

Brick proportions

Figure 4.2 indicates the basic relationship between header and stretcher faces with joint allowances for standard metric bricks. The cut pieces of whole bricks are used in bonding.

Purposes of bonding

The reasons for bonding brickwork are:

(i)   to strengthen a wall;
(ii) to ensure that any loads are distributed; and
(iii) to make sure that it is able to resist sideways or lateral pressure, see Fig. 4.3.

The straight joints of an unbonded wall make it weak and liable to failure, as shown in Fig. 4.4.

Principles of bonding

To maintain strength, bricks must be lapped one over the other in successive courses along the wall and in its thickness. There are two practical methods, using either a half-brick lap or a quarter-brick lap, called half-bond and quarter-bond (Fig. 4.5). If the lap is greater or smaller than these, then both appearance and strength are affected. If bricks are so

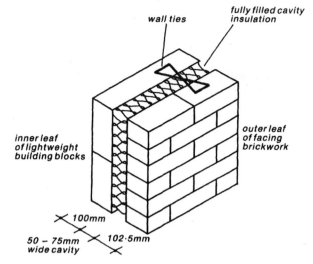

Figure 4.1   A typical external cavity wall

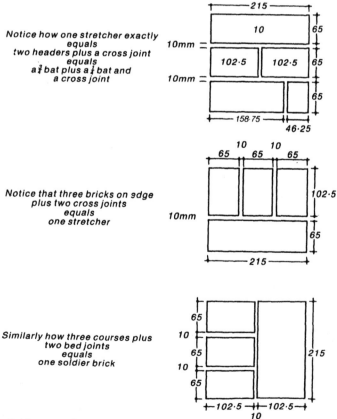

Notice how one stretcher exactly equals two headers plus a cross joint equals a ¾ bat plus a ¼ bat and a cross joint

Notice that three bricks on edge plus two cross joints equals one stretcher

Similarly how three courses plus two bed joints equals one soldier brick

Figure 4.2   Brick proportions

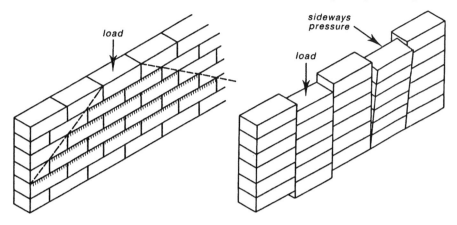

Figure 4.3 Bonding the bricks distributes any applied loads evenly

Figure 4.4 Tendency to failure of unbonded walls

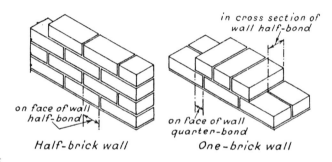

Figure 4.5

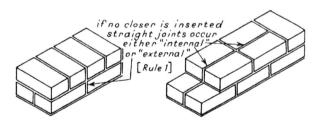

Figure 4.6

placed that no lap occurs, then the cross-joints or perpends are directly over each other (Fig. 4.6), and this is termed a 'straight joint,' being either 'external' for those appearing on the face of the wall, or 'internal' for those occurring inside the wall, and they should be avoided whenever possible. The apprentice should note that internal straight joints will occur in some bonding problems; excessive cutting would perhaps solve a particular problem, but this wastes labour and materials and tends to weaken the wall. On the other hand, by introducing one or two straight internal joints, whole bricks can be used. This is a case where practice and theory must compromise.

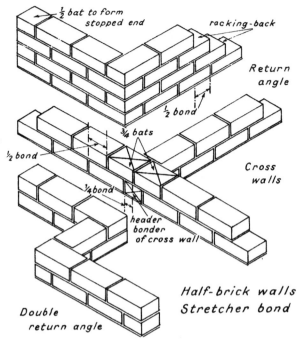

Figure 4.7 Junctions in half brick walls

The pattern in a brick wall is purposely arranged, has its particular use, and is called a 'bond.'

To summarise, the two main principles of the bonding of brickwork are:

1. To maintain half- or quarter-bond, avoiding at all times external straight joints and internal straight joints wherever possible.

2. To show the maximum amount of specified face bond pattern as possible.

To assist in maintaining these principles, rules should be remembered and applied (see Fig. 4.10).

The apprentice should never try to remember all the problems shown as examples. Problems must be solved as they occur by the logical application of the Rules. Eventually, the bonding of brickwork becomes automatic to the bricklayer.

Several bonds are in general use, but for the purpose of beginning the apprentice's bonding education, 'stretcher,' 'English,' and 'Flemish' bonds will be explained. Problems in other bonds can be solved by the application of the same rules.

Note. Wall thicknesses are usually stated in brick sizes, e.g. the width of a brick is known as a half-brick wall; the length of a brick as a one-brick wall; the width, plus length of a brick, as a one-and-a-half brick wall, and so on.

Stretcher bond: used in the building of half-brick walls. The face bonding consists entirely of stretchers, except where return angles, stopped ends, and cross walls occur. To gain maximum strength, half-bond must be maintained at all times. At the junction between two walls, quarter-bond is

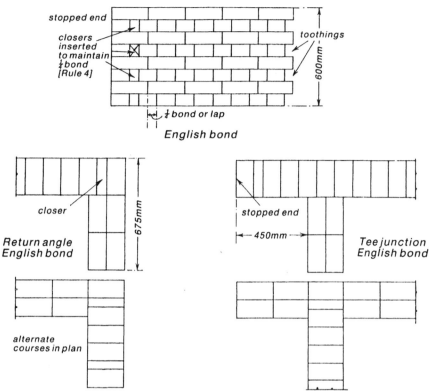

Figure 4.8 English bond

introduced, but by the insertion of three-quarter bats the wall can be continued in half-bond (Fig. 4.7).

English bond: alternate courses of headers and stretchers. A very strong bond, with no straight joints occurring in any part of the wall. Being monotonous in appearance, it is used in walls where strength is preferable to appearance; see Fig. 4.8.

Flemish bond: alternate headers and stretchers in the same course. Used in brick walls of a decorative nature. Internal straight joints, quarter-brick in length, occur at 100mm intervals along the middle of the wall. The header must be in the centre of stretchers in courses above and below (Fig. 4.9).

Rules of bonding

When working out bonding arrangements, the bricklayer makes use of a number of Rules of Bonding, which are used as a guide. Like all good rules, there are permitted exceptions.

Use whole bricks wherever possible; when the first two courses of Stretcher, English and Flemish bonds have been set out correctly they will repeat themselves and the perpends (vertical joints) in every other course will be upright or 'plumb.' The bricklayer checks this by plumbing perpends at every 900mm or so along the wall, course by course, see Fig. 4.17.

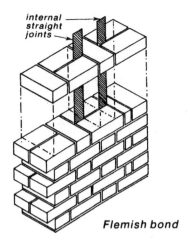

internal
straight
joints

Flemish bond

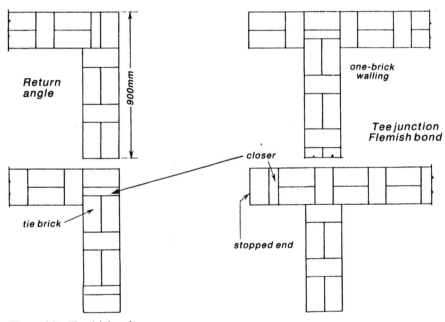

Return
angle

900mm

closer

tie brick

one-brick
walling

Tee junction
Flemish bond

stopped end

Figure 4.9   Flemish bond

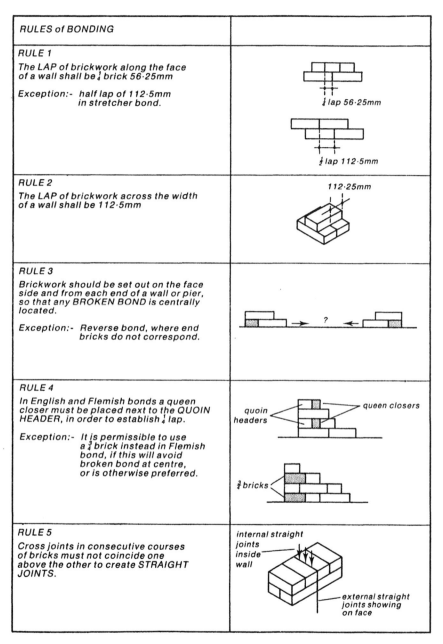

RULES of BONDING	
RULE 1 The LAP of brickwork along the face of a wall shall be ¼ brick 56·25mm Exception:- half lap of 112·5mm in stretcher bond.	¼ lap 56·25mm ½ lap 112·5mm
RULE 2 The LAP of brickwork across the width of a wall shall be 112·5mm	112·25mm
RULE 3 Brickwork should be set out on the face side and from each end of a wall or pier, so that any BROKEN BOND is centrally located. Exception:- Reverse bond, where end bricks do not correspond.	?
RULE 4 In English and Flemish bonds a queen closer must be placed next to the QUOIN HEADER, in order to establish ¼ lap. Exception:- It is permissible to use a ¾ brick instead in Flemish bond, if this will avoid broken bond at centre, or is otherwise preferred.	quoin headers · queen closers ¾ bricks
RULE 5 Cross joints in consecutive courses of bricks must not coincide one above the other to create STRAIGHT JOINTS.	internal straight joints inside wall external straight joints showing on face

Figure 4.10   Rules of bonding

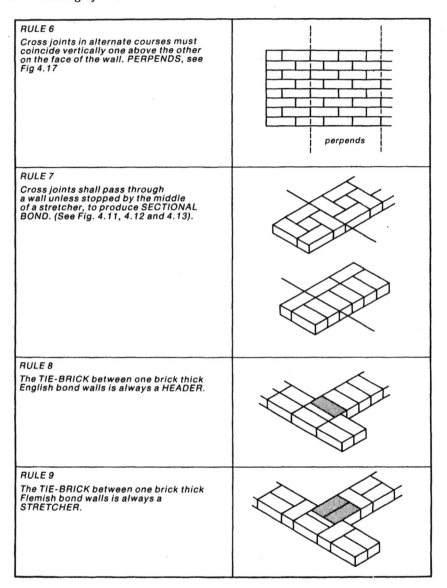

RULE 6

Cross joints in alternate courses must coincide vertically one above the other on the face of the wall. PERPENDS, see Fig 4.17

perpends

RULE 7

Cross joints shall pass through a wall unless stopped by the middle of a stretcher, to produce SECTIONAL BOND. (See Fig. 4.11, 4.12 and 4.13).

RULE 8

The TIE-BRICK between one brick thick English bond walls is always a HEADER.

RULE 9

The TIE-BRICK between one brick thick Flemish bond walls is always a STRETCHER.

Figure 4.10 cont.   Rules of bonding   (see page 44 also)

RULE 10 *In English bond, when a wall changes direction, the face bond changes from headers to stretchers or vice-versa in the same course. (See Fig 4.14).* *Exception:- This rule will not apply when (a) walls of different thickness intersect, and (b) walling curves on plan. (c) where the change of direction is 112·5mm break or less. (See Fig 4.16).*	*exception*
RULE 11 *The middle of thicker brick walls must be filled in with headers – to satisfy Rule 2. (See Fig 4.13 and 4.23).* *Exception:- half bricks are used in 1½ brick thick Flemish bond walling.*	Plan view — English bond Plan view — Flemish bond
RULE 12 *English bond walls which are an EVEN number of ½ bricks in thickness, will show the same bond on opposite faces of each course.* *English bond walls which are an ODD number of ½ bricks in thickness, will show stretchers on one side and headers the other, in each course.*	*four half bricks thick* *three half bricks thick*

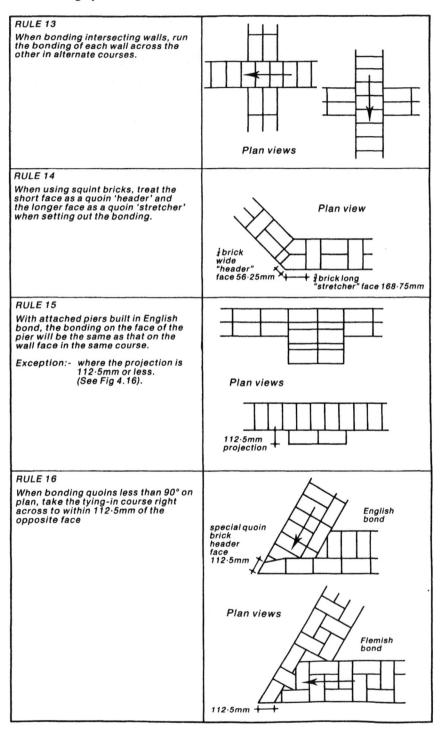

RULE 13

When bonding intersecting walls, run the bonding of each wall across the other in alternate courses.

Plan views

RULE 14

When using squint bricks, treat the short face as a quoin 'header' and the longer face as a quoin 'stretcher' when setting out the bonding.

Plan view

½ brick
wide
"header"
face 56·25mm

½ brick long
"stretcher" face 168·75mm

RULE 15

With attached piers built in English bond, the bonding on the face of the pier will be the same as that on the wall face in the same course.

Exception:- where the projection is 112·5mm or less. (See Fig 4.16).

Plan views

112·5mm
projection

RULE 16

When bonding quoins less than 90° on plan, take the tying-in course right across to within 112·5mm of the opposite face

special quoin
brick
header
face
112·5mm

English
bond

Plan views

Flemish
bond

112·5mm

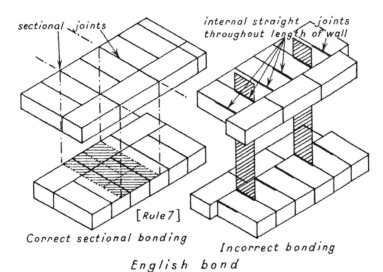

sectional joints

internal straight joints
throughout length of wall

[Rule 7]

Correct sectional bonding

Incorrect bonding

English bond

Figure 4.11

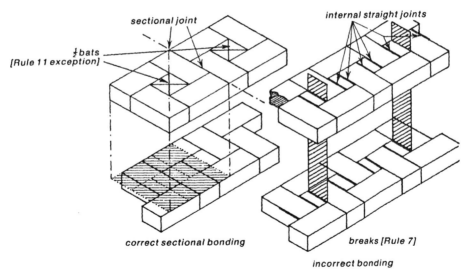

sectional joint

internal straight joints

½ bats
[Rule 11 exception]

correct sectional bonding

breaks [Rule 7]

incorrect bonding

Flemish bond

Figure 4.12

Illustrations of
the rules
of bonding

(continued on pages 46 to 48)

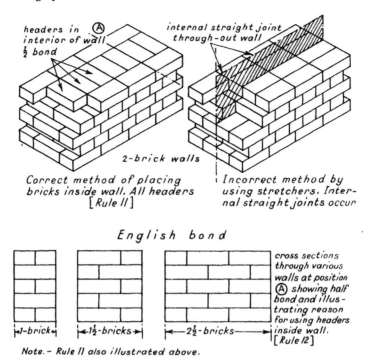

headers in Ⓐ interior of wall ½ bond

internal straight joint through-out wall

2-brick walls

Correct method of placing bricks inside wall. All headers [Rule 11]

Incorrect method by using stretchers. Internal straight joints occur

English bond

cross sections through various walls at position Ⓐ showing half bond and illustrating reason for using headers inside wall. [Rule 12]

|←1-brick→| |←1½-bricks→| |←2½-bricks→|

Note.- Rule 11 also illustrated above.

Figure 4.13

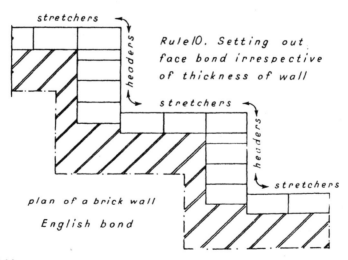

stretchers

headers

Rule 10. Setting out face bond irrespective of thickness of wall

stretchers

headers

stretchers

plan of a brick wall

English bond

Figure 4.14

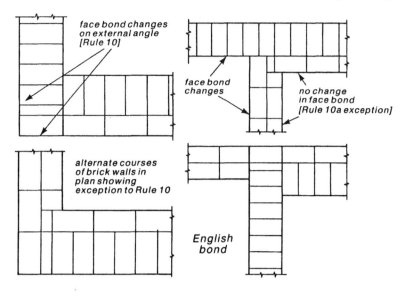

face bond changes
on external angle
[Rule 10]

face bond
changes

no change
in face bond
[Rule 10a exception]

alternate courses
of brick walls in
plan showing
exception to Rule 10

English
bond

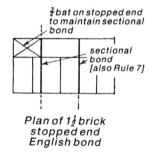

¾ bat on stopped end
to maintain sectional
bond

sectional
bond
[also Rule 7]

Plan of 1½ brick
stopped end
English bond

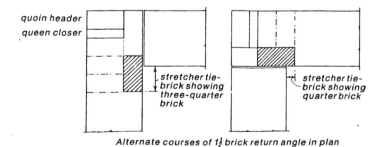

quoin header
queen closer

stretcher tie-
brick showing
three-quarter
brick

stretcher tie-
brick showing
quarter brick

Alternate courses of 1½ brick return angle in plan

Figure 4.15

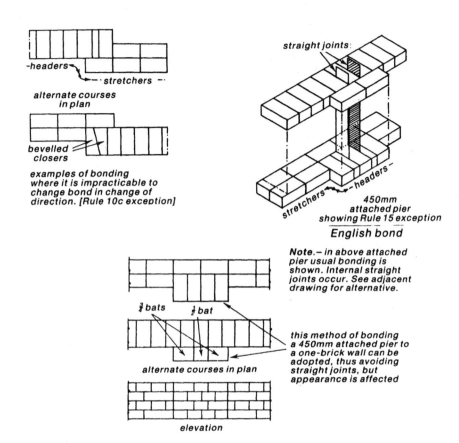

headers

stretchers

alternate courses
in plan

bevelled
closers

examples of bonding
where it is impracticable to
change bond in change of
direction. [Rule 10c exception]

straight joints

stretchers   headers

450mm
attached pier
showing Rule 15 exception

English bond

Note.– in above attached
pier usual bonding is
shown. Internal straight
joints occur. See adjacent
drawing for alternative.

¾ bats   ½ bat

alternate courses in plan

elevation

this method of bonding
a 450mm attached pier to
a one-brick wall can be
adopted, thus avoiding
straight joints, but
appearance is affected

Figure 4.16

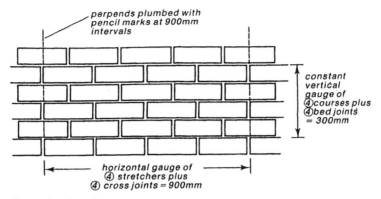

Stretcher bond

Coordinating dimensions for brickwork which take into account variations in size of bricks. The same principles shown here on an elevation of stretcher bond apply to other face bonds

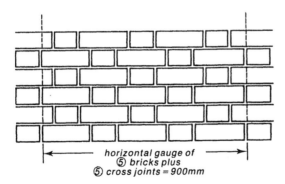

Flemish bond

Figure 4.17 Plumbing perpends

Problem 1

Set out in plan the alternate courses of a 1½-brick return angle in English bond.

First draw the outline of the problem to be solved (Fig. 4.18), then set out the first course on the external or internal angle, whichever is selected as face side. In ½-brick and 1-brick walls only one face side is obtainable. Walls thicker than one brick have two face sides, the thickness of the wall joint being adjustable to allow for the variation in brick size. Setting out in this case begins on the more important side.

Beginning this problem with the external angle—Rules 4 and 10 apply (Fig. 4.19). Continue by placing the tie-brick—Rule 7 (Fig. 4.20). Complete 'backing in'—Rule 7—and, on stopped end use a ¾ bat (Fig. 4.21). If no stopped end occurs or the wall is not of this particular length, i.e. 900mm, the tie-brick can be arranged as in Fig. 4.22.

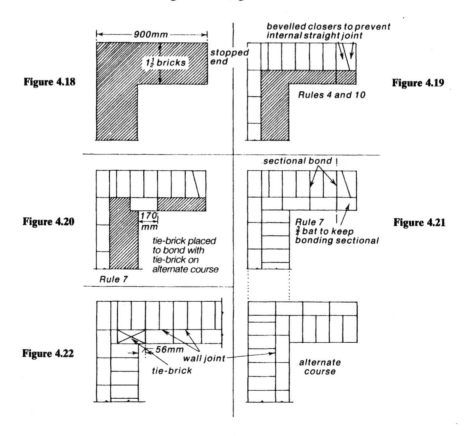

Figure 4.18

Figure 4.19

Figure 4.20

Figure 4.21

Figure 4.22

Problem 2

Set out in plan the alternate courses of a two-brick return angle in English bond (Fig. 4.23).

Outline the problem to be solved.

Set out the external face side, applying Rules 4 and 10. Continue by setting tie-brick, and 'backing-in'—Rules 7, 10 and 12. Fill in the interior of the wall—Rule 11.

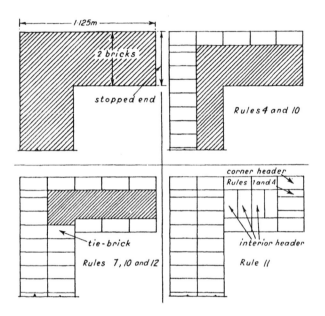

Figure 4.23

Problem 3

Set out in plan the alternate courses of a 1½-brick return angle in Flemish bond.

Outline the problem to be solved (Fig. 4.24).

Set out the external face side, applying Rules 4 and 7 (Fig. 4.25).

Place the tie-brick—Rule 7—(Fig. 4.26), and complete problem.

Note the two headers together on the internal face side (Fig. 4.27); this is termed 'broken bond,' and the position shown is the most usual and the best for this particular problem. The apprentice will find it is possible to place the broken bond nearer the angle and in some bonding problems this arrangement would probably be more suitable.

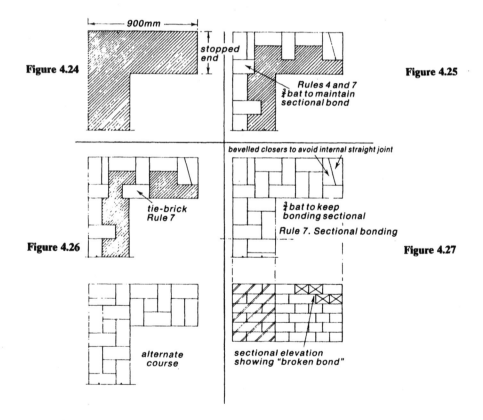

Figure 4.24

900mm

stopped end

Figure 4.25

Rules 4 and 7
¾ bat to maintain
sectional bond

Figure 4.26

tie-brick
Rule 7

bevelled closers to avoid internal straight joint

¾ bat to keep
bonding sectional

Rule 7. Sectional bonding

Figure 4.27

alternate
course

sectional elevation
showing "broken bond"

Understanding brick dimensions _____

The *co-ordinating size* of clay bricks, inclusive of mortar joints, is given in BS3921:1985 as 225mm long by 112.5mm wide by 75mm deep, see Fig. 2.4.

The *work size*, for which brick manufacturers aim (also given in the BS) is 10mm less than each of those dimensions, to provide a 'joint allowance'. The work size for each standard metric brick is therefore 215mm long by 102.5mm wide by 65mm deep.

The *actual sizes* of bricks as delivered vary slightly, and these variations must be absorbed by adjusting the thickness of mortar joints.

The individual differences in brick dimensions (larger or smaller than the work size) are mainly due to differing rates of drying and shrinkage of the naturally occurring clay or shale—the raw material. This is where the skill of the bricklayer is important. Opening or tightening up cross joints must be carried out carefully and evenly, so that:

(i)   it is not noticeable in the finished wall;
(ii) plumb perpends are maintained at three or four brick intervals for the full height of the wall;
(iii) the co-ordinating size of the brickwork is maintained.

Figure 4.17 shows how varying sizes of brick must be kept within vertical and horizontal gauge.

Broken bond _____

This occurs where the length of a wall or pier, from end to end, does not work out to suit the bond pattern exactly.

In accordance with Rule 3, any broken bond should be located towards the middle of a wall or pier. Face work should be set out allowing a 225mm space for each stretcher with its cross joint. Similarly, allowing 112.5mm for each header plus joint.

On long walls there may be scope for slight adjustment of cross-joint thickness, from this basic module of 225mm, in order to avoid broken bond and so achieve economy and a good appearance. (For example, tightening up 1mm on every cross joint in the first course of a stretcher bond wall 8.95m long would allow 40 whole bricks to be fitted into that length, using 9mm rather than 10mm joints).

Broken bond can arise for two reasons.

(i)   Where the architect's overall dimension of a wall does not equal a number of whole bricks. The cut portion of a brick located near the middle of a wall must never be smaller than a half bat, that is 102.5mm, see Fig. 4.28.
(ii) Where the wall length does not need any cut bricks at the centre, but is still not a convenient length to maintain the correct repetition of bond pattern, see Fig. 4.29.

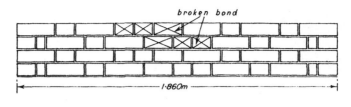

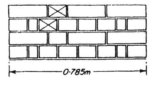

Figure 4.28 Location of broken bonding (inconvenient dimensions—solved using cut bricks)

Examples of broken bond

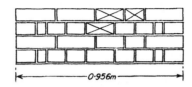

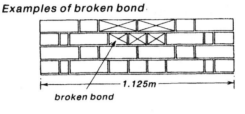

Figure 4.29 Location of broken bonding (using whole bricks)

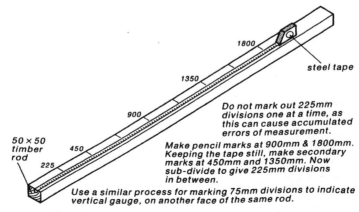

Do not mark out 225mm divisions one at a time, as this can cause accumulated errors of measurement.

Make pencil marks at 900mm & 1800mm. Keeping the tape still, make secondary marks at 450mm and 1350mm. Now sub-divide to give 225mm divisions in between.

Use a similar process for marking 75mm divisions to indicate vertical gauge, on another face of the same rod.

Figure 4.30 Marking out a gauge or storey rod

For whichever reason broken bond applies, once located in the first two courses of a wall, the perpends of this broken bond must be plumbed carefully for the *full* height of the building elevation.

Setting-out facework in a wall without openings

Before setting-out the bond on the face side of any wall, it is wise to make a 'gauge rod'. This should be made of straight, smooth timber 50mm × 50mm in cross section, and approximately 3m long. Along one face fine sawcuts carefully made at 225mm intervals will allow stretcher bond to be set out consistently from end to end, leaving any broken bond near the middle of the wall.

On another surface of the rod, sawcuts at 75mm intervals serve to check regular vertical gauge when bricklaying commences and quoins are raised. Use of this dual-purpose 'storey rod', for checking horizontal and vertical gauge, avoids potential risk of error when using a measuring tape.

Setting out a facework wall which includes openings

The principle here is to preplan window and door openings *before* a face brick is bedded. It is no good raising a facework wall to window sill level, then asking 'Right, now where do the windows go?'

If you do this, you will not get continuous perpends, straight and plumb, from top to bottom of the wall! First, mark out the location and widths of window and door openings by measurement on the ground beam or substructure brickwork, see Fig. 4.31.

Providing architect's chosen dimensions A to E are multiples of 225mm, then no broken bond will occur anywhere with stretcher and English bonding. Different dimensions for A to E will operate for Flemish or other bonding patterns, if broken bonding is to be avoided

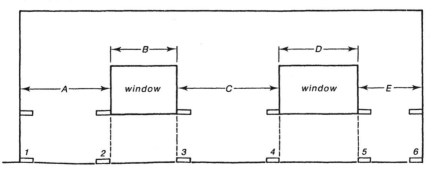

Elevation of facebrick walling

Reveal-brick stretchers 1 to 6 identified at ground level before facework commences. These become firm perpend plumb points for full height of walling.

Figure 4.31 Setting out facework

'Reveal bricks' should be placed dry as indicated in Fig. 4.31. The bond pattern is then laid out to left and right of these fixed-point reveal bricks.

Keeping to the co-ordinating dimension of 225mm horizontal gauge, any broken bond will be chased along to an approximately central point between the reveal bricks.

This system will locate any broken bond centrally under windows and/or in the middle of window piers.

When the perpends of each reveal brick are plumbed, as the wall is raised up, there will be a whole brick conveniently in place to form the vertical sides of every window opening when sill level is reached. Figure 4.32 shows some further variations when setting openings in facework.

Reducing the visual effects of broken bond

Broken bond in a facework wall can be visually disturbing above or below each window. The effect can be minimised if the mortar joint colour matches the brick colour. Light coloured mortar with dark bricks makes broken bond very obvious. This is not the bricklayer's fault, as broken bond is not incorporated by choice. Broken bonding costs time and money, both to set out and to cut the necessary bricks for the full height of the walling. If the widths of window and door openings are made to suit brick sizes *and* the chosen bond pattern, then there will be no broken bond at all. Figure 4.33 indicates a method of making broken bond less obvious in a stretcher bond wall.

Reverse bond

Occasionally it is possible, when setting out facework, to avoid broken bond at the centre of a wall or window pier by 'reversing' the bond. This means operating the exception to Rule 3. Experience tells the foreman bricklayer where and when it is possible to use this. Figure 4.34 gives three examples where reverse bond has been used as an alternative to broken bond.

It is not advisable to make use of reverse bond when setting out facework if a building has contrasting-coloured facing bricks ('dressings') at the sides of window and door openings. This would upset the balance of appearance, see Fig. 4.35.

Bonding examples

From the information already given in this chapter, the apprentice will readily understand the illustrated examples which follow. These will serve to introduce some craft terms, in addition to showing some of the problems which confront a bricklayer in practice.

An offset in the plan view of a wall may be referred to as a double return, because there are two quoins or returns formed, see Fig. 4.7. Alternatively,

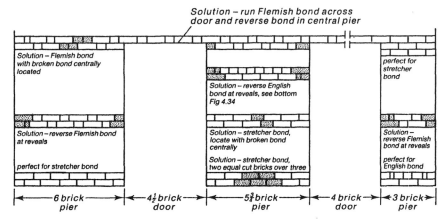

Solution – run Flemish bond across
door and reverse bond in central pier

Solution – Flemish bond
with broken bond centrally
located

Solution – reverse English
bond at reveals, see bottom
Fig 4.34

perfect for
stretcher
bond

Solution – reverse Flemish bond
at reveals

perfect for stretcher bond

Solution – stretcher bond,
locate with broken bond
centrally

Solution – stretcher bond,
two equal cut bricks over three

Solution –
reverse Flemish
bond at reveals

perfect for
English bond

|←——6 brick——→|←—4½ brick—→|←——5¾ brick——→|←—4 brick—→|←3 brick→|
| pier | door | pier | door | pier |

*once set out at ground level, perpends of reveals & broken bond must be plumbed
for the full height of the walling*

Figure 4.32 Setting out facework, various bonds. Further examples of using broken bond, or
avoiding it with reverse bond

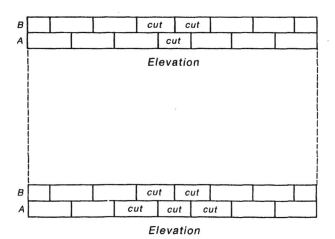

B | | | cut | cut | | |
A | | | cut | | |

Elevation

B | | | cut | cut | | |
A | | cut | cut | cut | |

Elevation

*If the **single** cut brick in course 'A' of the upper panel is
turned into three **equal** size cut bricks in course 'A' of
the lower panel, then the broken bonding is much less obvious*

Figure 4.33 Broken bond in stretcher bond walling

it may be called a 'break' in the line of the wall. See typical bonding
arrangements in Fig. 4.36.

Note the two methods of dealing with a ½ brick recess reveal in Flemish
bond. There is less cutting in 'A' (Fig. 4.37) and it is probably the most
practical. In any case, the cutting of reveal bricks is always difficult,
especially in the case of hand-made and engineering bricks, the former
creating excessive waste and the latter being almost impossible to cut with
hammer and bolster.

The best method is to use purpose-made bricks (see Fig. 13.4, BD.1,

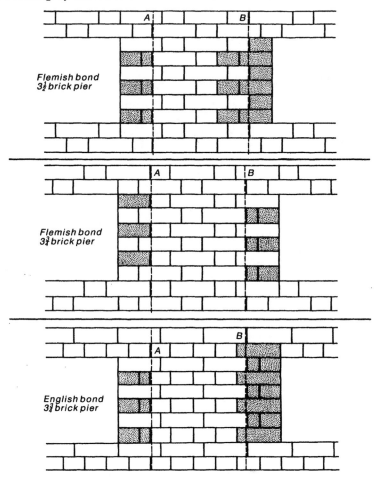

Notice how the firm perpends A and B (dotted) continue unbroken for the full
height of the elevation. This fundamental principle is most important,
particularly with light-colour mortar and dark-colour bricks and vice versa.

Note also how, in each case, the reverse bond (shaded) disappears above
and below window openings, unlike setting out broken bond, which will be
continuous from top to bottom of wall elevation

Figure 4.34   Three examples of facework, where reverse bond has been used to avoid broken
bond in window piers

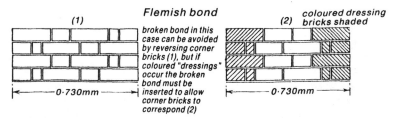

Figure 4.35   Broken bond may be inevitable, particularly where contrast colour bricks are
required at window reveals

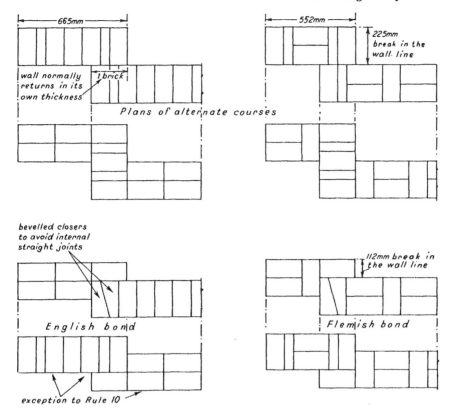

Figure 4.36 Bonding double returns, or breaks, in English and Flemish bond

BD.2 and BD.3). If these are not available, see Fig. 4.38 for hand cutting by hammer and bolster.

Figure 4.39 show two methods of dealing with the same problem. Number 2 is the better method, cut bricks being avoided on the face of the brick wall.

Figures 4.40 to 4.42 show attached piers. Note the arrangement of bricks in bonding a 330mm attached pier in Flemish bond (Fig. 4.40 lower half). The appearance of the attached pier is English bond, nevertheless it is stronger and looks better than the method sometimes used (Fig. 4.41), which is often mistakenly described as Flemish bond, but which is, in fact, half-bonding.

Figure 4.42 shows two ways of arranging a 440mm attached pier in Flemish bond. The top method uses a broken bond on the straight face; the other avoids this.

Figures 4.43 and 4.44 illustrate junctions in 1½-brick walling. Note the two arrangements of bricks in junctions.

Figures 4.45 to 4.47 show a brick wall introducing a 1½-brick double return angle and two methods of finishing a stopped end in a 1½-brick wall.

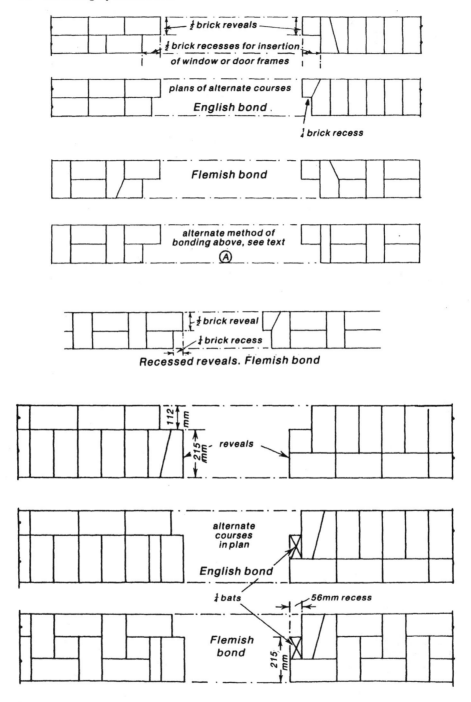

Figure 4.37   Bonding recessed reveals

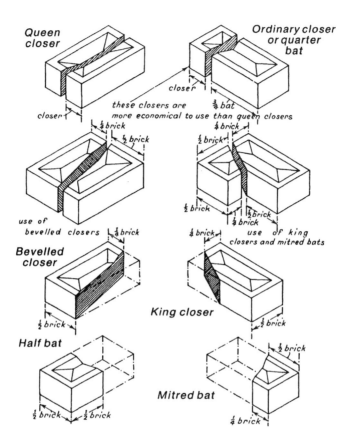

Figure 4.38 Definitions of cut bricks

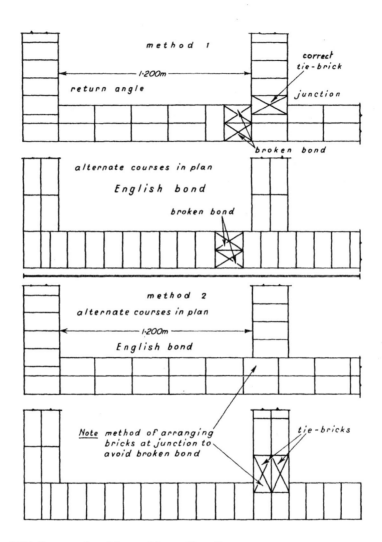

Figure 4.39   Economy in solving problems of bonding

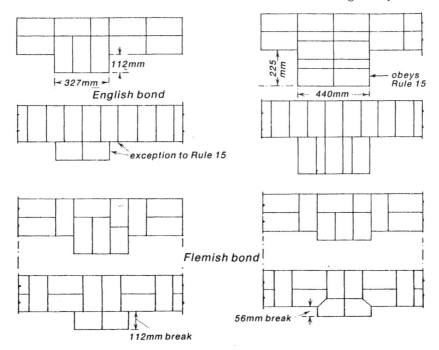

English bond

Flemish bond

112mm

327mm

225 mm

440mm

obeys Rule 15

exception to Rule 15

112mm break

56mm break

Figure 4.40 Attached piers

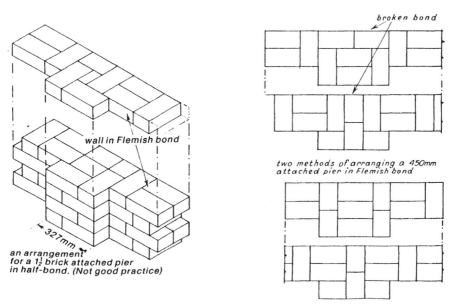

wall in Flemish bond

327mm

an arrangement for a 1½ brick attached pier in half-bond. (Not good practice)

broken bond

two methods of arranging a 450mm attached pier in Flemish bond

Figure 4.41 The stretcher bond on the face of the pier does not match the Flemish bond walling behind

Figure 4.42

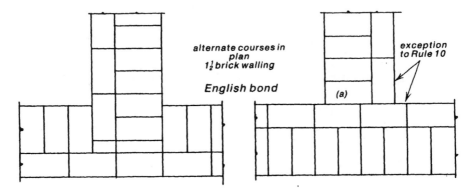

Figure 4.43   Junction walls

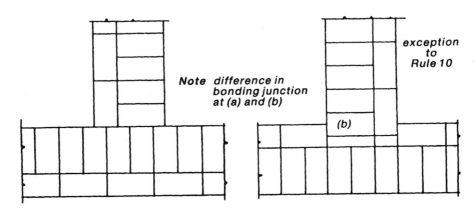

Figure 4.44   Junction walls—alternative bonding

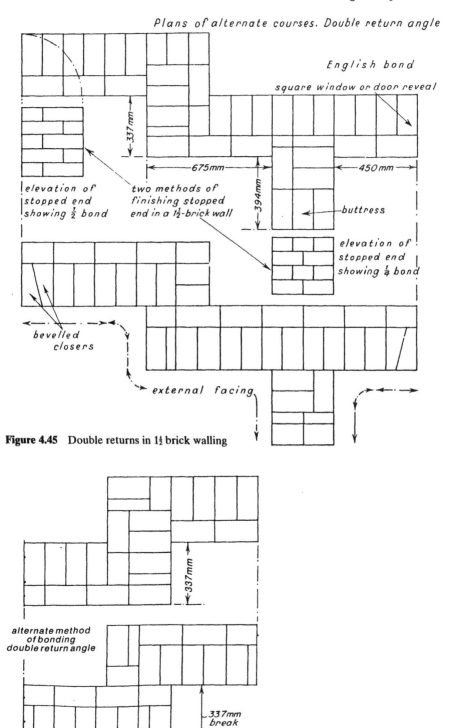

Figure 4.45 Double returns in 1½ brick walling

Figure 4.46 Alternative to Fig. 4.45

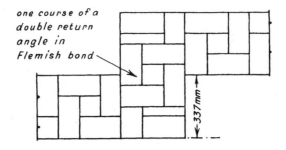

one course of a
double return
angle in
Flemish bond

337mm

Figure 4.47

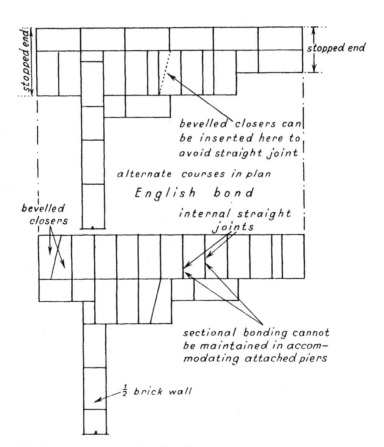

stopped end

stopped end

bevelled closers can
be inserted here to
avoid straight joint

alternate courses in plan

E n g l i s h   b o n d

bevelled
closers

internal straight
joints

sectional bonding cannot
be maintained in accom-
modating attached piers

$\frac{1}{2}$ brick wall

Figure 4.48   An interesting example of bonding

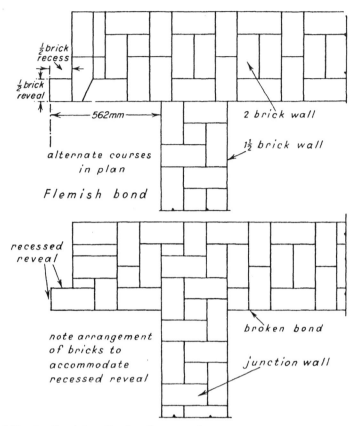

½ brick recess

½ brick reveal

|←———562mm———→|

alternate courses in plan

Flemish bond

2 brick wall

1½ brick wall

recessed reveal

note arrangement of bricks to accommodate recessed reveal

broken bond

junction wall

Figure 4.49   Another interesting bonding example

It is usual, where stopped ends are to be rendered or partly hidden by a door or window frame, to build with the elevation showing half-bond. Less cutting is needed than in the case of the stopped end on buttress, where an elevation showing quarter-bond is necessary to maintain a correct appearance.

Figures 4.48 and 4.49 show two problems in which rules cannot be strictly adhered to. The figures are self-explanatory. Examine them carefully and set out the problem.

Summary _____

This chapter has set out the basic Rules of Bonding applicable to solid walling, giving examples of walls up to two bricks in thickness. Although a great deal of brickwork is in the form of cavity wall construction involving a half brick thick, stretcher bonded outer leaf only, it is important for the bricklayer to be confident with the Rules of Bonding as applied to walls of greater thickness than 102mm and to be familiar with other bond patterns.

The next chapter introduces other face bonds and applications of bonding details.

5
Bonding details

This chapter deals with bonding details involving the use of bricks of special shape called plinths, and of squints, and where walls do not intersect at right angles, e.g. obtuse angle quoins of more than 90° and of acute angles of less than 90° in plan. (Chapter 13 describes the range of BS Special Shape bricks available).

The arrangement of face bonds other than English, Flemish and Stretcher bonds are illustrated, followed by coverage of isolated piers, bond patterns for decorative panels, and the formation of brickwork curved on plan.

Plinths

Figures 5.1(a) to (c) show bonding details where substructure brickwork is set back, or 'set in', (in this case by 28mm) for architectural effect, at the commencement of the superstructure, giving a building the appearance of a 'plinth' or base. In some buildings this 'plinth thickening' may be constructed from common bricks and be rendered with cement and sand mortar afterwards.

Purpose-made plinth bricks, with a 45° sloping surface, are used to achieve this set-on of face brickwork for a few courses above ground level, by 56mm per course. These plinth headers and stretchers are based upon the standard $215 \times 102 \times 65$mm size to fit in with bonding and gauge of main walling. An 'external return' special plinth is shown in Fig. 5.2(a).

Plinths are always a source of anxiety to the foreman bricklayer, as difficulties arise in maintaining correct bond on the face and at the same time avoiding unnecessary cuts and straight joints. Figure 5.1(a) shows a 28mm set-on acting as a plinth, and Fig. 5.1(c) is the arrangement of bricks at the quoin. This is the method usually adopted. An alternative is shown (Fig. 5.1(b)), where continuity of perpends is an absolute necessity. Figure 5.2(a) should be carefully noted.

Preplan the bonding

The bonding arrangement for any plinth brickwork must be preplanned well in advance of starting work. A little time spent with squared paper, drawing it out beforehand, will avoid the embarrassment of straight joints later on.

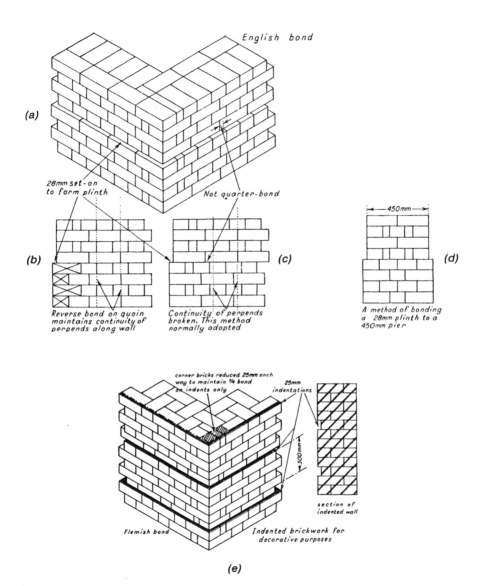

Figure 5.1 Indented brickwork for decorative purposes

70 Bonding details

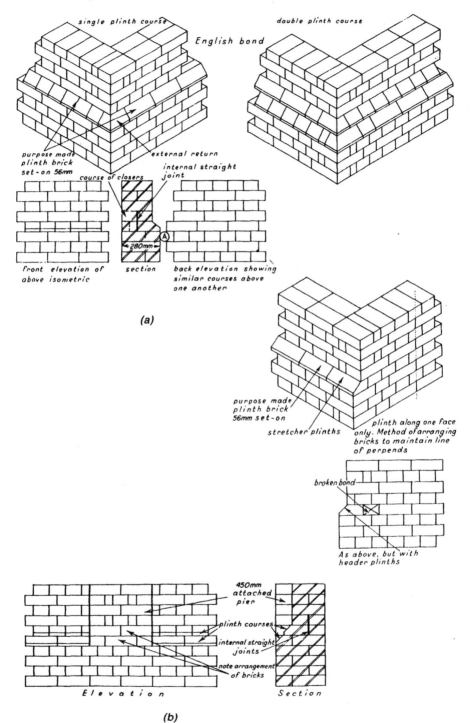

Figure 5.2 Plinths

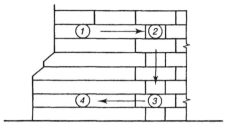

Figure 5.3 Preplanning bonding procedure for plinth brickwork

Draw in the face bonding required on squared paper in the numerical order shown in Fig. 5.3. Start above the plinth courses on both faces of a quoin. Plumb bonding perpends down through the plinth courses. Adjust the face bonding from (3) towards (4) in Fig. 5.3, to avoid straight joints on both faces of the quoin, using broken bond and bevelled closers.

Figure 5.3 if carefully followed, should enable an apprentice to set out most plinth problems. Figure 5.2(b) shows the setting out of an attached pier above plinth courses. The arrangement of bricks is sound and avoids unnecessary cutting.

Plinth bricks may also be used upside down to construct the brick corbel courses shown in Fig. 7.36 (page 117), to give a more decorative effect.

Indented brickwork

Indenting one course along a whole elevation, as shown in Fig. 5.1(e), produces interesting shadow effects. Alternatively, this indenting can be restricted to approximately $1\frac{1}{2}$ bricks only in each direction at quoins, for the same purpose.

Decorative brick quoins

Figure 5.4 shows other ways of making the angles of a building look more obvious; this is sometimes described as 'rusticated' blocked corners and dressed corners.

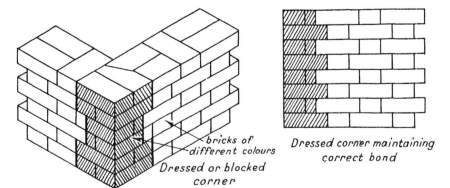

bricks of different colours

Dressed or blocked corner

Dressed corner maintaining correct bond

Figure 5.4 Contrast colour bricks used at quoins

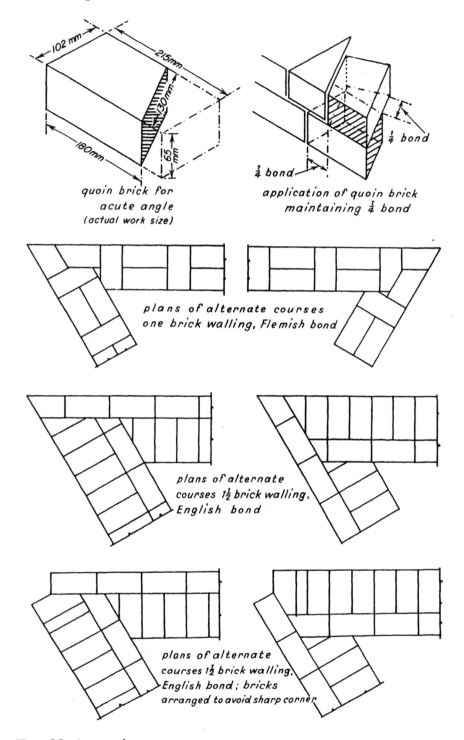

quoin brick for
acute angle
(actual work size)

application of quoin brick
maintaining ¼ bond

¼ bond

¼ bond

plans of alternate courses
one brick walling, Flemish bond

plans of alternate
courses 1½ brick walling,
English bond

plans of alternate
courses 1½ brick walling,
English bond; bricks
arranged to avoid sharp corner

Figure 5.5   Acute angles

Acute angles

(Figure 5.5). Note the method of planning quoin brick to obtain ¼-bond along wall. Owing to its small size, a closer is omitted from this corner; it would be difficult to cut and awkward when laying.

Obtuse angles or squint corners

(Figure 5.6). Note the planning of quoin brick. Both acute and obtuse angle bricks are planned to conform to standard brick size and are most generally used. Others which are used are normally over-size and purpose-made to the architect's specification. Note the alternate courses of brickwork at 'birdsmouth' angles, 'A' in the bottom element of Fig. 5.6. Careful craftsmanship is needed here to avoid the appearance of a straight joint, by emphasizing the tie bricks to left and right when jointing up.

Panel walls and isolated piers

(Figures 5.7 and 5.8). The panel walls are planned to show a broken bond. Note the latter's position in this case. The 330mm isolated pier 'A' avoids straight joints but requires an excessive number of cut bricks. Pier 'B' has internal straight joints and shows half-bonding on two sides but is considered sound enough for most purposes and is therefore more generally adopted.

Other types of bond

Single Flemish bond

(Figure 5.9). This is a combination of English and Flemish bonds. One side of the wall shows Flemish bond and the other English, which being stronger is given preference in the thickness of the wall. Flemish bond is therefore only a facing and is bonded in by its headers on every other course. Single Flemish bond is used:

    1. For its combination of strength and appearance.

    2. For economy of facing bricks, 'snap' headers being used on every other course.

    3. If the architect wishes to show a different bond on each side of the wall to conform with the type of brick in use.

Dutch bond   (Figures 5.10 and 5.11). Dutch bond is similar to English bond in appearance, the difference being in the external corner. A three-quarter bat takes the place of the header and closer, and a half-bat or header is introduced into every other stretcher course next to the corner three-quarter.

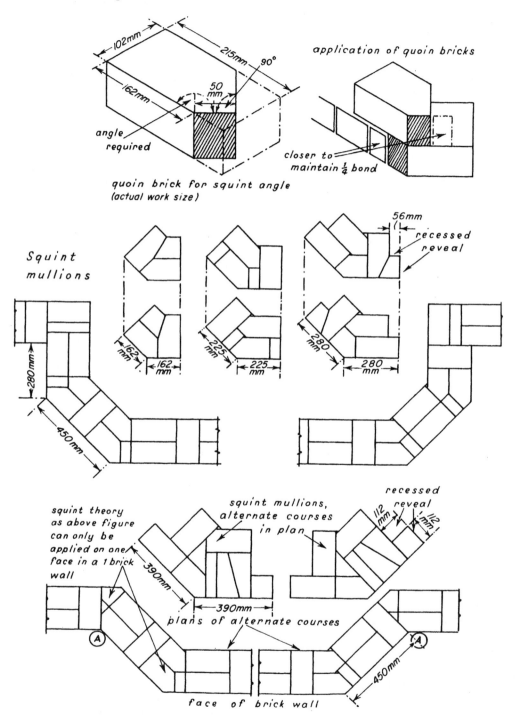

Figure 5.6 Obtuse or squint angles

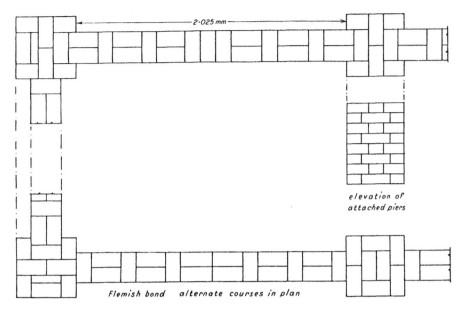

Figure 5.7 Panel walls

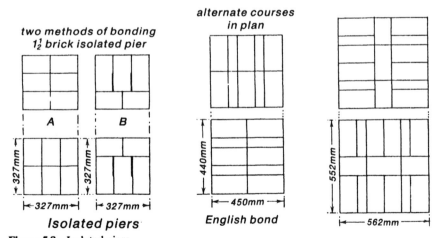

Figure 5.8 Isolated piers

The advantages of this bond are:

1. Bricks of a different colour from those being used for the main brickwork can easily be introduced into the wall face, forming patterns termed 'diapers', see Figs. 5.10 and 7.45.

2. Loads are more evenly distributed owing to the fact that regular quarter-brick offsets occur in the bond.

3. When using a multi-coloured brick, an attractive wave effect is obtained.

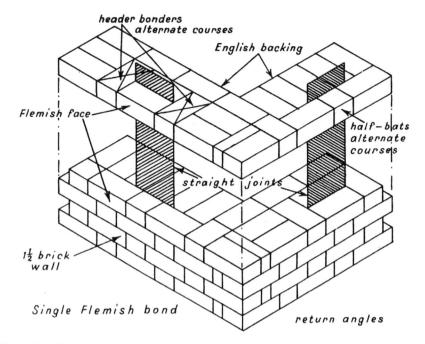

header bonders
alternate courses

English backing

Flemish face

half-blats
alternate
courses

straight joints

1½ brick
wall

Single Flemish bond

return angles

Figure 5.9   Single Flemish bond

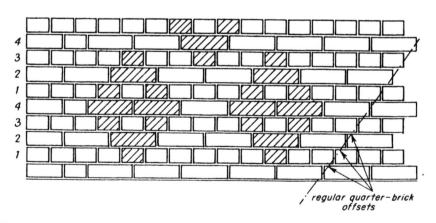

regular quarter-brick
offsets

Figure 5.10   Dutch bond

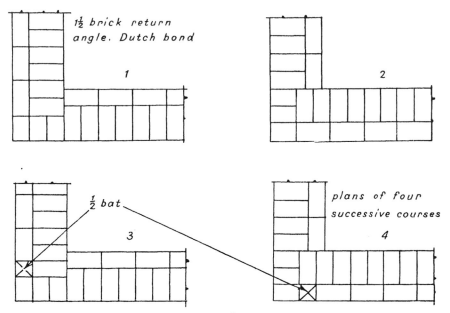

Figure 5.11 Dutch bond example applied to 1½ brick walling

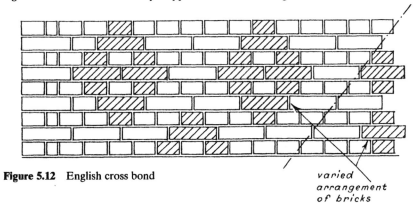

Figure 5.12 English cross bond

Figure 5.13 Monk bond

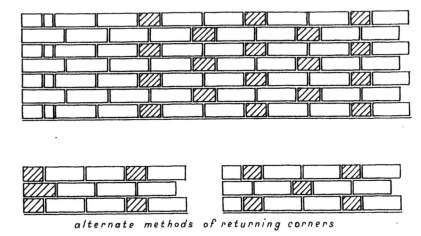

alternate methods of returning corners

Figure 5.14   Variations of monk bond

(note the half-bonding of the stretcher courses)

Figure 5.15   English garden wall bond

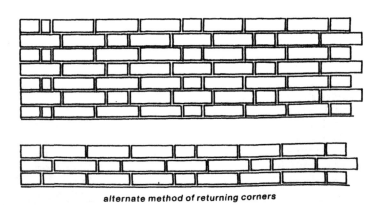

alternate method of returning corners

Figure 5.16   Flemish garden wall bond

English cross bond: similar to Dutch bond, except that a header and closer are introduced in place of a three-quarter corner brick (Fig. 5.12).

Monk bond: two stretchers to one header in the same course. This facing bond introduces a zig-zag effect into the wall. Note that the bricks forming the pattern are exactly 1½ bricks apart as shown in Fig. 5.13. If this is specially noted, no difficulty will be experienced in maintaining the correct bond. For variations of Monk bond or of '2 and 1' bond see Fig. 5.14.

Garden wall bonds: facing bonds in which many more stretchers than headers are used. Several internal straight joints occur, but the advantages are:
  1. Fair faces can be obtained on both sides of a one-brick wall.
  2. Economises in facing bricks.

English garden wall bond consists of three courses of stretchers to one of headers (Fig. 5.15). An internal straight joint occurs throughout the length of the wall, three courses in height.

Flemish garden wall bond consists of three stretchers to one header in the same course. To maintain the correct bond, the header in one course must be in the centre of the middle stretchers in courses above and below (Fig. 5.16). In contrast to normal English and Flemish bond this bond is stronger than English garden wall bond, the header bonders being more evenly distributed.

Rat-trap bond: bricks laid on edge in form of Flemish bond. Used for garden walls or where walls of a building are to be vertically tiled, or as it is termed 'tile-hung.' The pockets formed inside the wall can be left hollow or made solid according to instructions (Fig. 5.17).

Quetta bond: Used in walls 1½-bricks or more in thickness, which are reinforced vertically with concrete and mild steel rods. The face bond is in either Flemish or Flemish garden wall bond. It will be noted that small reinforced concrete columns are formed at intervals along the wall (Fig. 5.18).

Single herring-bone, double herring-bone, basket-weave and diagonal basket-weave bonds (Figure 5.19): decorative bonds used in panel work for floors and paths or vertically in walls. Note the method of setting out the herring-bone bonds. This ensures cut bricks of the same size on all sides of the panel. For a small square or rectangular panel the setting out must commence from the centre of panel. For a long narrow panel, the setting out can begin from the bottom centre. Many architects design decorative panels by combining these bonds, by introducing 'creasing' tiles, or by other arrangements of bricks.

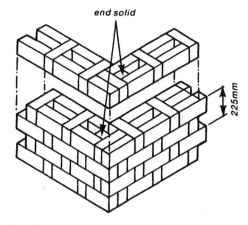

Figure 5.17   Rat trap bond

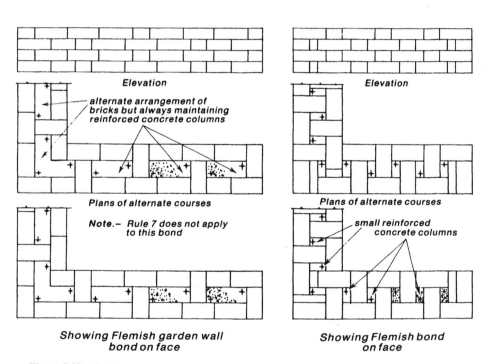

Figure 5.18   Quetta bond

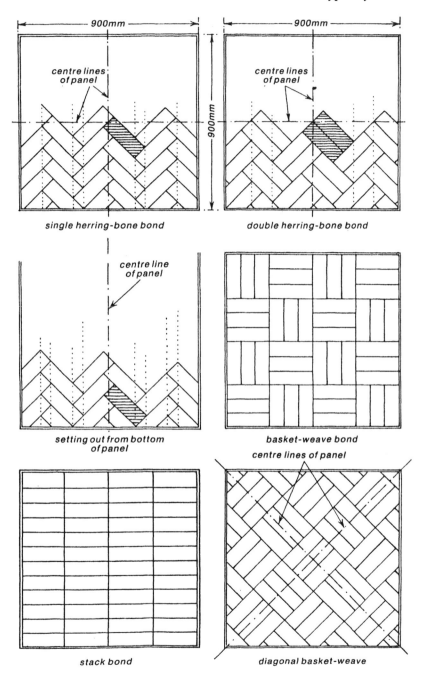

Figure 5.19   Decorative panels of brickwork

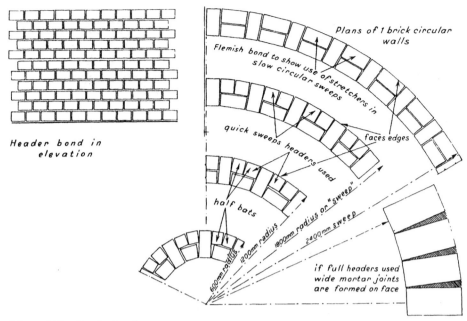

Figure 5.20   Header bond

Header bond (Figure 5.20): used in the construction of circular brickwork. In the building of circular cesspools and similar work where a small radius occurs, purpose-made radiating bricks are preferable (see Fig. 13.4 RD.1 and RD.2), but in general construction the ordinary brick can be used. To avoid excessive cutting, the bond is arranged as shown in the figure, i.e. two half-bats and a one-brick bonder placed alternately. If uncut bricks are used, unduly wide cross-joints would appear on one side of the wall (see the bottom right-hand element in Fig. 5.20).

With the use of either Commons or hand-made facing bricks, header bond is suitable, up to a radius of approximately 2250mm. Beyond this measurement, the normal type of bond can be introduced. Where a glazed or similar brick is being used, it is necessary to extend the radius to approximately 4000mm before adopting normal bond. The brick, being regular in shape and size, tends to 'hatch and grin' (see Fig. 7.3(a)) if stretchers are used on smaller radius.

The apprentice will encounter other kinds of bonds, but these are merely variations of those already mentioned. They have no specific name but are used by the architect when considered more suitable to use with the type of brick that has been specified.

6

Foundations

The ground or subsoil on which a building rests is called the 'natural foundation' or 'sub-foundation,' and has a definite load-bearing capacity, according to the nature of the soil. Subsoils are of many varieties and may generally be classified as rock, compact gravel or sand, firm clay and firm sandy clay, silty sand and loose clayey sand. It is possible to erect a wall on rock with little or no preparation, but on all soils it is necessary to place a continuous layer of *in-situ* concrete in the trench called the 'building foundation'. (The term *'in situ'* means cast in place in its permanent position; unlike 'precast' which is made elsewhere, lifted and transported later to the place where it is required for use). This cast, *in-situ* concrete is made from Portland cement with coarse aggregate, plus sharp sand or ballast, graded from 40mm to fine sand, and mixed in the proportion of 1:6.

The design of a foundation is an important subject for the architect's consideration, particularly where the building structure is heavy, with possible concentrated loading, and the ground on which the building rests is of poor load-bearing capacity or is affected by other conditions such as seasonal change. Even buildings of the small domestic and industrial type may require careful site exploration so that a suitable foundation is constructed and the future stability of the building assured. Where buildings are extremely heavy it may be necessary to construct foundations of a special type, such as 'piles,' where the site consists of deep beds of soft soil overlaying a hard soil, but it is the intention here to deal only with buildings of the smaller domestic and industrial range.

Standard foundations

When a load is placed on soil it is necessary to 'spread' or 'extend' the foundation base in order to ensure stability. This extended or spread foundation is referred to as a 'strip foundation' in the case of a continuous wall structure, or a 'pad foundation' in the case of an isolated pier.

The important loads to be resisted by a foundation are the 'dead loads,' or the weight of the building structure, and the superimposed loads, or the weight that may be placed on the building structure. The area or spread of the foundation base should be sufficient to resist the downward thrust or bearing pressure of these combined loads. The spread of a pad foundation (Fig. 6.1) required for a pillar or isolated pier can be determined by

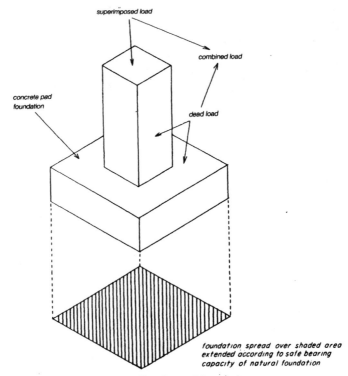

Figure 6.1 Transfer of building load on to subsoil under pad foundation

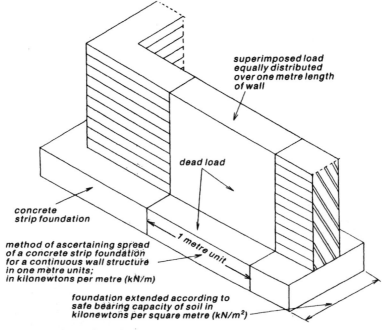

Figure 6.2 Loading under strip foundation

dividing the combined loads by the safe bearing capacity of the soil, viz:

$$\frac{\text{Combined load of pier}}{\text{Safe bearing capacity of soil per square metre}} \quad \text{or} \quad \frac{\text{Total load of pier}}{\text{Safe load on soil}}$$

The foundation for a continuous wall structure (Fig. 6.2) is obtained by similar methods; in this case a one metre length of wall is taken for the purpose of calculation.

Having determined the spread of the foundation base it is now necessary to consider the depth. The base, in spreading the load, is subjected to stresses known as 'tension' and 'punching shear' and must therefore be of sufficient depth to resist them. Tension is due to the bending tendency and punching shear to the possibility of the wall or pier structure's tendency to punch a hole through the foundation base (Fig. 6.3). One method of ascertaining the depth of concrete is shown in Fig. 6.4.

Where the structural loads are very heavy or where the safe bearing capacity of the soil is low, the spread of the foundation base would become greater and this is normally referred to as a 'wide strip foundation.' It follows that the required depth of the concrete foundation base may be considered excessive and it can be reduced by the introduction of steel reinforcement, but the foundation must always be of sufficient depth to ensure that, in combination with the steel, it will resist the stresses of tension and shear. Fig. 6.5 shows a simple example of a reinforced concrete strip foundation.

Loads are not always evenly distributed along the wall, but may at times be concentrated at various points. Fig. 6.6 shows part of a wall with an attached pier, which may, for instance, bear a roof truss or beam. In order to obtain even pressure over the soil the foundation would be extended as shown.

The depth below ground level to which foundations should be taken depends upon the type of soil and is the depth at which the sub-foundation ceases to be affected by the weather. This is between 600mm and 1500mm, decreasing as the proportion of gravel increases, (Fig. 6.4).

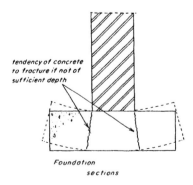

tendency of concrete to fracture if not of sufficient depth

Foundation sections

Figure 6.3 Potential failure of strip foundation if concrete is too thin

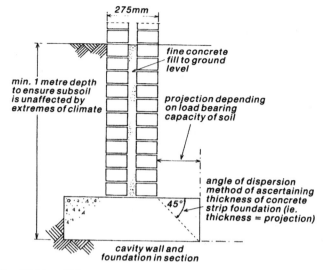

275mm

fine concrete
fill to ground
level

min. 1 metre depth
to ensure subsoil
is unaffected by
extremes of climate

projection depending
on load bearing
capacity of soil

angle of dispersion
method of ascertaining
thickness of concrete
strip foundation (ie.
thickness = projection)

45°

cavity wall and
foundation in section

Figure 6.4 Establishing thickness of strip foundation concrete

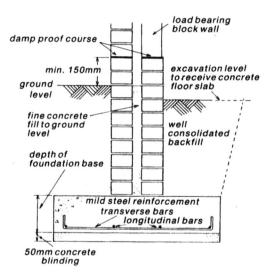

load bearing
block wall

damp proof course

min. 150mm

ground
level

excavation level
to receive concrete
floor slab

fine concrete
fill to ground
level

well
consolidated
backfill

depth of
foundation base

mild steel reinforcement
transverse bars
longitudinal bars

50mm concrete
blinding

Figure 6.5 Reinforced concrete wide strip foundation

In buildings of the smaller domestic and industrial type, or where the loads are not excessive, standard methods and rules are applied to the design of the foundations. Generally, the resultant construction complies with the Building Regulations, but in all cases the decision as to adequacy rests with the local authority building control office.

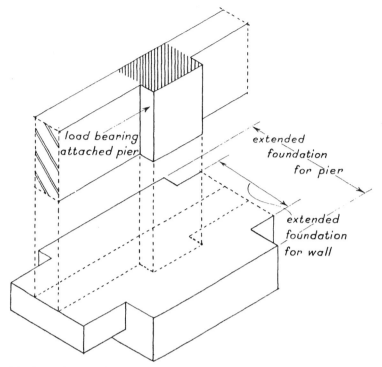

Figure 6.6 Extension of strip concrete foundation around attached pier

Brick footings _____

Due mainly to the quality of concrete now used in modern construction practice, many authorities consider brick footings to be obsolete. However, it is felt that the apprentice should be aware of the principles involved when this form of construction is adopted and of the necessary bonding arrangements.

It is possible to ease the stresses of tension and shear in a foundation base by the addition of a construction known as a 'footing' and where brick is used this is achieved by regular offsets at the base of the wall or pier structure; the footings spread the weight of the wall or pier and super-imposed loads over the concrete, which in turn distributes the combined loads over the soil.

Thus the first course of footings is always *double* the wall width, the second and each following course of footings is offset 56mm each side. Every course of footings should be header bond and sectional, following Rule of Bonding 7, see Fig. 6.7.

Stepped foundations

These are constructed on sloping sites in order to ensure a horizontal bearing on the natural foundations (Fig. 6.8). Note the overlap of concrete

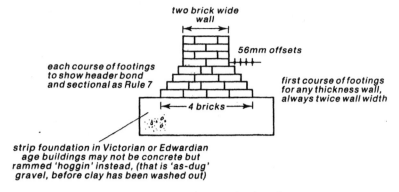

two brick wide
wall

56mm offsets

each course of footings
to show header bond
and sectional as Rule 7

first course of footings
for any thickness wall,
always twice wall width

4 bricks

strip foundation in Victorian or Edwardian
age buildings may not be concrete but
rammed 'hoggin' instead, (that is 'as-dug'
gravel, before clay has been washed out)

Figure 6.7 Cross section of footing courses to a two-brick wall

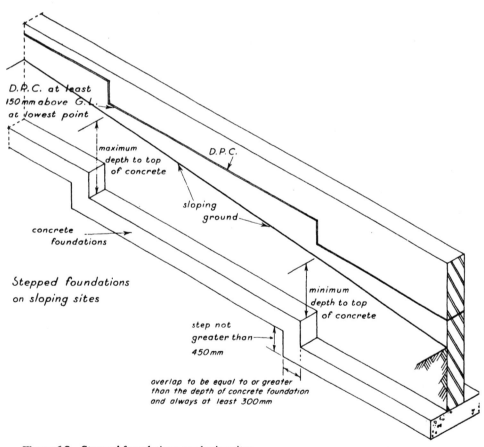

D.P.C. at least
150mm above G.L.
at lowest point

maximum
depth to top
of concrete

D.P.C.

sloping
ground

concrete
foundations

Stepped foundations
on sloping sites

minimum
depth to top
of concrete

step not
greater than
450mm

overlap to be equal to or greater
than the depth of concrete foundation
and always at least 300mm

Figure 6.8 Stepped foundations on sloping sites

at the change of levels. The height of the steps must be maintained at not more than 450mm and where this dimension is exceeded special precautions may be necessary.

Special foundations

Standard strip, see Fig. 6.4 (or trench-fill, see Fig. 9.6) foundations are suitable for most of the buildings under discussion, but the apprentice may find variations. Sub-foundations of made-up ground or of shrinkable clays, for instance, require the careful consideration of the architect and structural engineer, who may decide to adopt a simple reinforced concrete raft foundation or a system of short-bored piles and ground beams.

7

Craft operations

Bricklaying

Good bricklaying entails the ability to master the art of spreading the mortar bed, dexterity in handling the brick to be laid, and the possession of a keen eye. All these can be acquired by practice. Before laying bricks on any job, place the mortar or 'spot' board, with the bricks, in a convenient position. They must be within easy reach so that no unnecessary movement is involved when materials are required. Block up the spot board on bricks, one at each corner, so that it is kept clean, and load out as shown (Fig. 7.1).

Do not grasp the trowel as if clenching the fist, but place the thumb on the ferrule and handle lightly, so that a flexible wrist action is possible (Fig. 7.2). Pick up the mortar with an easy sweeping motion and spread it on the wall sufficiently thick to allow the brick to be placed by pressure of the hand. A common fault is the placing of too much mortar under the brick, so that considerable hammering and tapping are necessary before the brick reaches its final position. The bricklayer usually estimates the amount of mortar bed required by the feel of the brick.

Bricks of the hand-made type are often slightly misshapen and some difficulty may be experienced in keeping a flat face and preventing 'hatching and grinning' (Fig. 7.3a). If bricks are cambered in their length, lay them as shown in Fig. 7.3b. Never allow them to 'cock up' at the back (Fig. 7.3c), as this makes the laying of the next course difficult and tends to place the wall out of level in its width. The method shown sometimes necessitates the laying of the brick frog-downwards and the filling of the frog with mortar before laying, to maintain the solidity of the wall. In ordinary circumstances, however, bricks should always be laid with the frog upwards.

Engineering bricks are always difficult to lay, as they have a tendency to 'swim' owing to their density and non-absorptive nature. When laying these bricks, see that the mortar is as stiff as possible for easy handling, lay the bricks in the required position and leave them, as any attempt to touch them after laying will create difficulties. Have confidence when handling them and do not resort to tricks such as sprinkling the mortar bed with neat cement or placing absorptive bricks on the mortar bed to withdraw the moisture before laying engineering bricks. Both are unnecessary—the former is costly, while the latter tends to destroy the hardening action of the cement mortar.

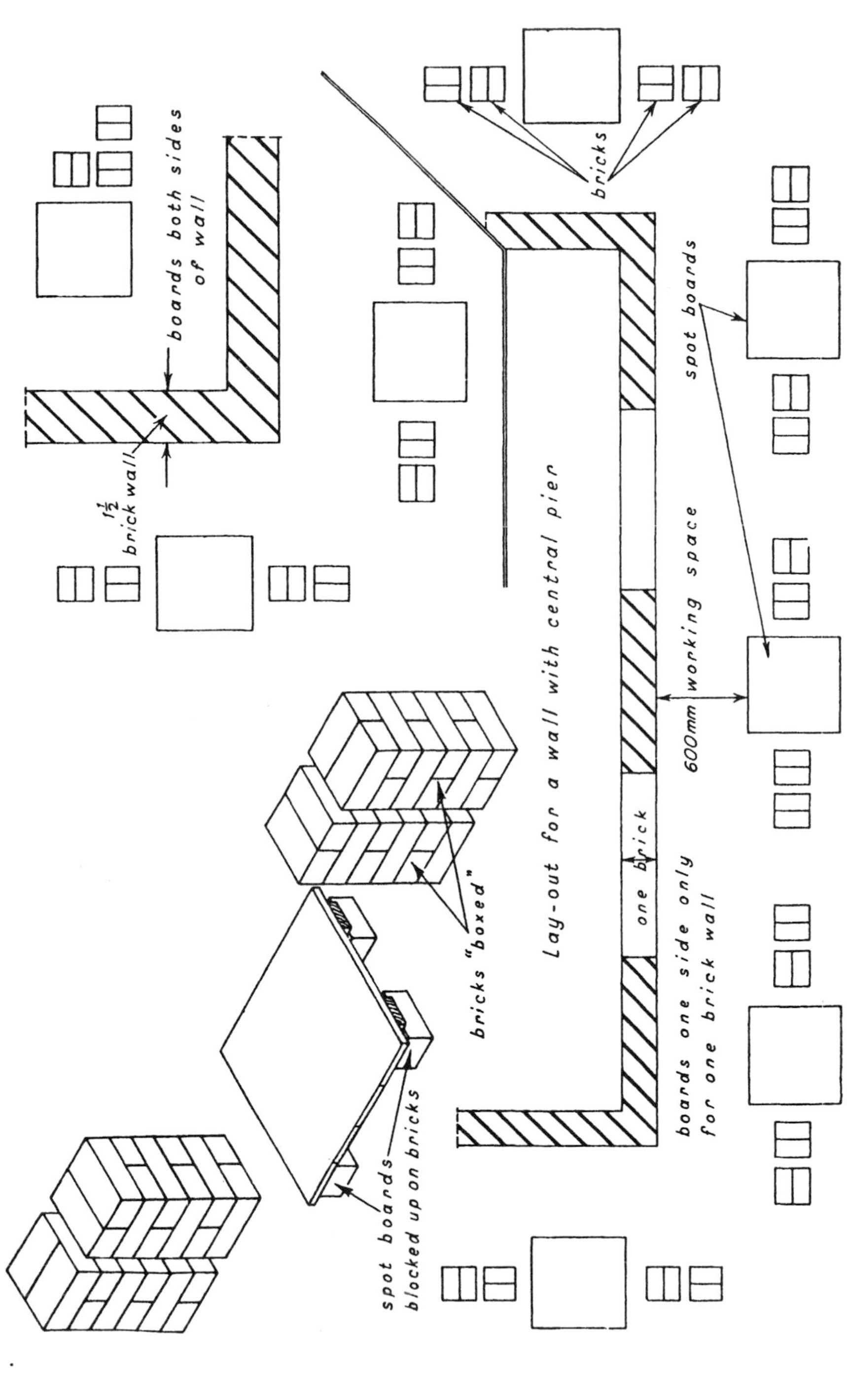
boards both sides
of wall
1½
brick wall
Lay-out for a wall with central pier
bricks
spot boards
600mm working space
one brick
boards one side only
for one brick wall
bricks "boxed"
spot boards
blocked up on bricks

Figure 7.1 Suggested layouts for spot boards and bricks in preparation for building

Figure 7.2   Method of holding brick trowel

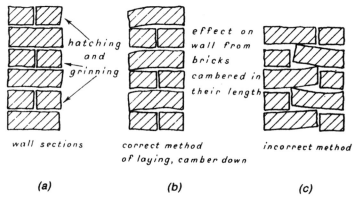

wall sections

correct method
of laying, camber down

incorrect method

(a)   (b)   (c)

Figure 7.3   Dealing with the natural characteristics of hand-made and stock bricks

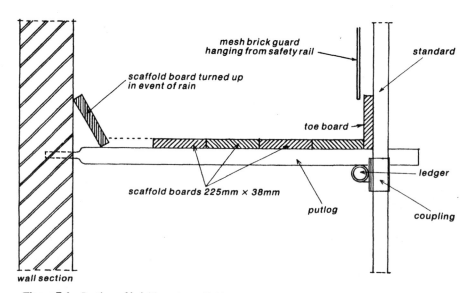

mesh brick guard
hanging from safety rail

standard

scaffold board turned up
in event of rain

toe board

scaffold boards 225mm × 38mm

putlog

ledger

coupling

wall section

Figure 7.4   Section of bricklayer's scaffold

In hot summer weather bricks, other than those of the engineering type, should be wetted to wash off surplus dust and to prevent undue absorption of moisture from the mortar bed. In winter months this is not necessary as the atmosphere is usually sufficiently damp to achieve these purposes. At this time of the year, however, brickwork should be protected overnight against frost. After the day's work is completed and before leaving the job, the last course of brickwork should be covered with sacking or other suitable material which may be available.

Face work, too, becomes stained through rain splashing back from the scaffold or from the platform from which the bricklayer works. This can be prevented by turning the front scaffold board on edge, as shown (Fig. 7.4).

Erecting a brick wall

When building a wall over 1.125 m in length it is always advisable to use a line and pins, for this will ensure a neat job. Before bringing these into use, it is first necessary to build the corners, making sure that they are vertical or 'plumb,' in alignment (Fig. 7.5), and to 'gauge.'

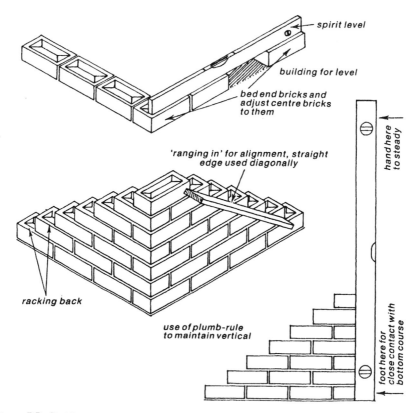

Figure 7.5 Raising quoins

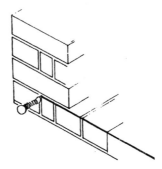

Figure 7.6   Use of line pins

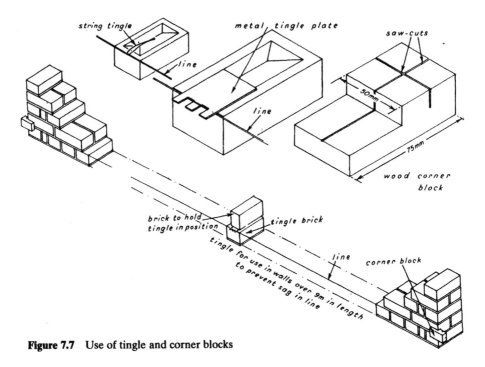

Figure 7.7   Use of tingle and corner blocks

If we assume the building of a straight one-brick wall, first, erect approximately six courses of brickwork on the corners and then begin to use a line and pins. Two ways of keeping the line taut are shown. Fig. 7.6 shows a method commonly used—the placing of the line pin in a vertical joint. Fig. 7.7, showing the use of corner blocks, illustrates the better method, especially where expensive facing bricks are being used.

When laying bricks to a line, always ensure that a trace of daylight can be seen between the line and brick (Fig. 7.8). This prevents the laying of the bricks 'hard' to the line, which if continued, would eventually place the wall out of alignment to a considerable extent. To keep the wall flat and to prevent hatching and grinning, imagine the bricks are being laid between *two* lines, one being the edge of the previous course laid to a line 'A' (Fig. 7.8) and the other to the present string line position at 'B'.

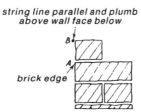

Figure 7.8   Laying to the line

If a wall exceeds 9 m in length, it is necessary to use a 'tingle' to take the sag out of the line. The tingle brick should be as near the centre of the wall as possible, and must be sighted through from corner to corner every time the line is raised one course, to ensure that the string line is being supported at the correct level (see the bottom right hand end of Fig. 7.7). A tingle achieves its purpose up to a wall length of approximately 12 m. Beyond this, it is advisable to divide the wall into two parts and to erect part of the wall in its centre to act as a corner. This should be located at a vertical movement joint or be beneath a pier, and is termed a 'lead' (Fig. 7.9). The wall illustrated is considered to have been erected, in order to show clearly the positions of the lead and tingle.

One of the duties of the foreman bricklayer is to see that sufficient work is in hand so that his bricklayers are continuously employed. To do this, he must have all corners erected in advance, in preparation for the building of the walls. Corners should never be built with a straight line of toothings, as in Fig. 7.10, as it is difficult to make good the toothings in all cases. The correct method is to build, as shown in Fig. 7.11, continually racking back and maintaining alignment of the corner by means of a profile. A combination of racking back and toothing is shown in Fig. 7.12. This prevents a direct line of toothings and avoids excessive racking out of brickwork.

Raising quoins

'Corner bricklayers' must always keep ahead of those running the line, so that there is always somewhere to fix line and pins or corner blocks throughout the day.

It is preferable if corner bricklayers continually raise small quoins, say, only six or seven courses ahead of the line, to avoid the quoin courses getting out of face plane alignment with the main walling.

Quoins must be large enough to resist the pulling power of the line each time it is tightened up. They must also be of just sufficient height to allow accurate ranging-in with the spirit level as indicated in Fig. 7.5.

However, if it is necessary for some other reason to build a large quoin, as shown in Fig. 7.11, than a temporary timber profile must be set up as indicated, so as to ensure that the racking-back courses will be truly in line with the overall wall face.

Fig. 7.13 shows the method of erecting a wall, with a number of attached piers. In order to maintain correct levels, it is always advisable to lay the course of brickwork on the wall before attempting to lay any bricks on the piers.

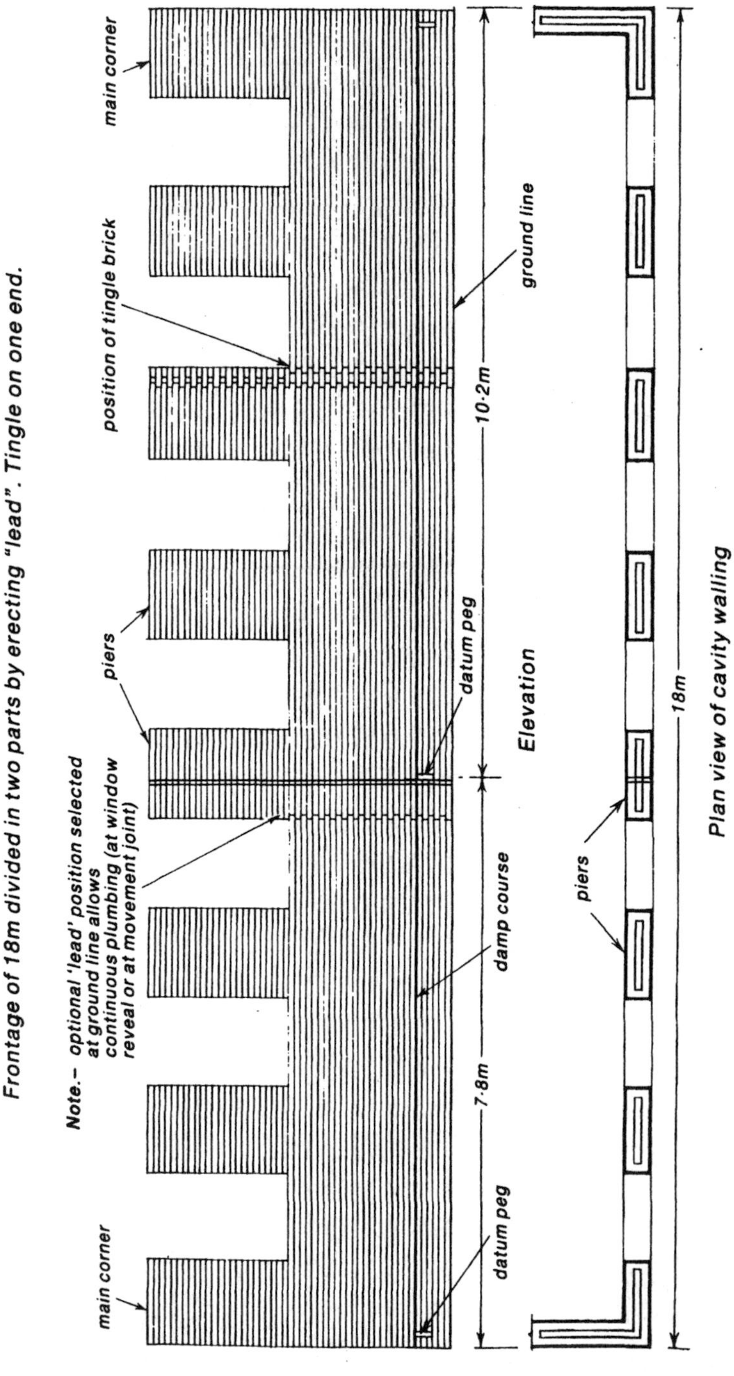

Frontage of 18m divided in two parts by erecting "lead". Tingle on one end.

main corner

position of tingle brick

piers

Note.– optional 'lead' position selected at ground line allows continuous plumbing (at window reveal or at movement joint)

datum peg

damp course

7·8m

datum peg

main corner

10·2m

ground line

Elevation

piers

18m

Plan view of cavity walling

Figure 7.9   Erecting a long wall

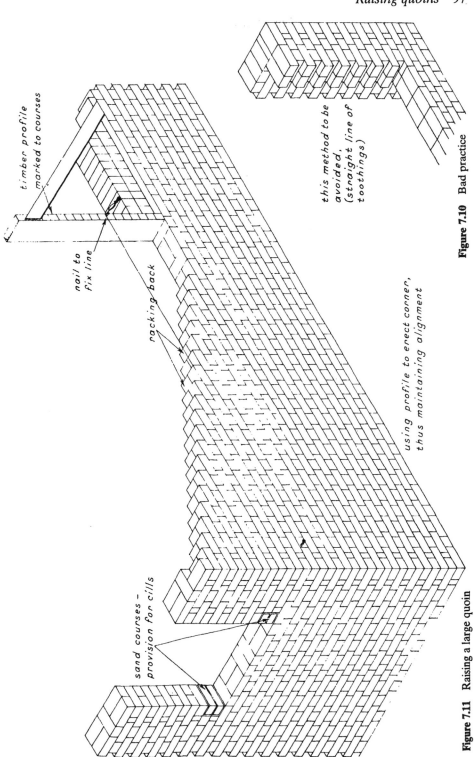

timber profile
marked to courses

nail to
fix line

racking-back

this method to be
avoided,
(straight line of
toothings)

Figure 7.10 Bad practice

using profile to erect corner,
thus maintaining alignment

sand courses –
provision for cills

Figure 7.11 Raising a large quoin

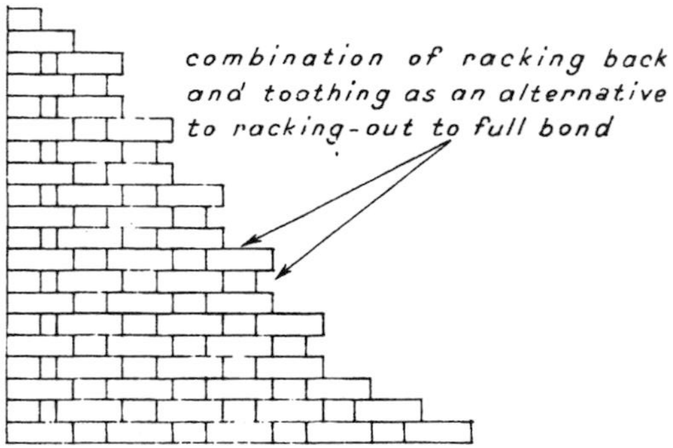

combination of racking back and toothing as an alternative to racking-out to full bond

Figure 7.12 . More convenient sized alternative to Fig. 7.11

Figure 7.13   Erecting a wall with attached piers

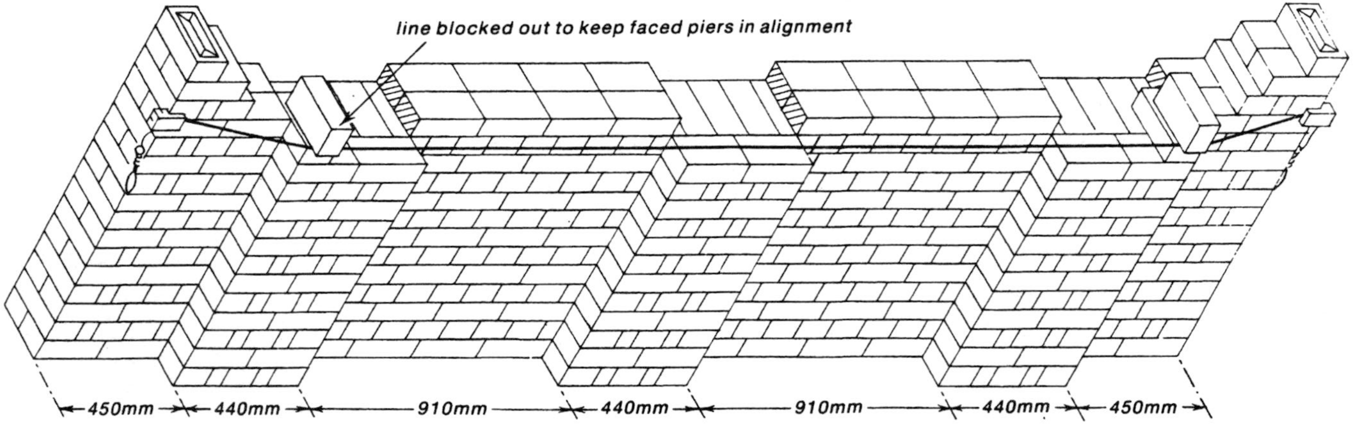

line blocked out to keep faced piers in alignment

450mm   440mm   910mm   440mm   910mm   440mm   450mm

Note: Actual dimensions of attached piers and intermediate spaces are shown, which will apply when setting out instead of nominal dimensions of 450mm and 900mm respectively

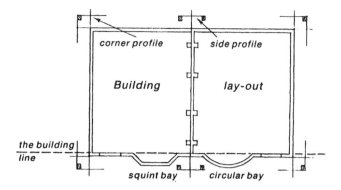

Figure 7.14   Location of foundation profiles

Preliminary setting-out of a building ‾‾‾‾‾

It is assumed that the building of a traditional semi-detached domestic dwelling is contemplated (Fig. 7.14), but it should be noted that the same principles can be applied to the terrace and crosswall types of construction and other more complicated forms. The frontage line of the building is ascertained either from a measurement taken from the centre of the road or from a building opposite. Corner profiles are placed to this frontage line, the main post of one profile being set to a given level (usually representing the ground-floor level), called the 'datum,' and being concreted to prevent movement. The back profiles are set square to the frontage line by means of a large builder's square with sides approximately 2 m to 3 m and they are checked by the application of the right-angled triangle theory, known to the bricklayer as the '3, 4, 5 method' (Fig. 7.15).

Side profiles for centre walls are set into position and are marked with concrete and wall lines, nails being used to assist in the adjustment of ranging lines.

Framed timber moulds must be prepared for the setting out of brick bays. They are applied as shown in Fig. 7.16.

Two methods can be adopted for the construction of circular bays:

1. The use of a framed timber mould for original setting out, and circular templets for building. This is the usual method, especially where a number of bays are to be built (Fig. 7.17).

2. The use of a trammel (Fig. 7.18).

The procedure for construction to the damp course level will be as follows:

1. The vegetable soil is removed from the site.

2. The ranging lines are set to concrete marks on the profiles and trenches are excavated. To prevent the sides of excavations from collapsing, a temporary framed timber structure ('timbering') is used. Some simple forms for shallow trench excavations are shown in Fig. 7.19. After the trenches have been excavated, the concrete foundations are placed in them.

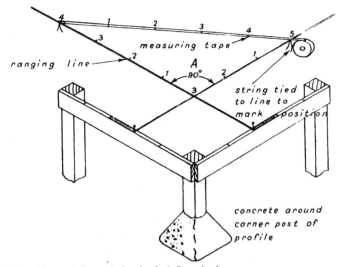

Figure 7.15   Checking a right angle by the 3–4–5 method

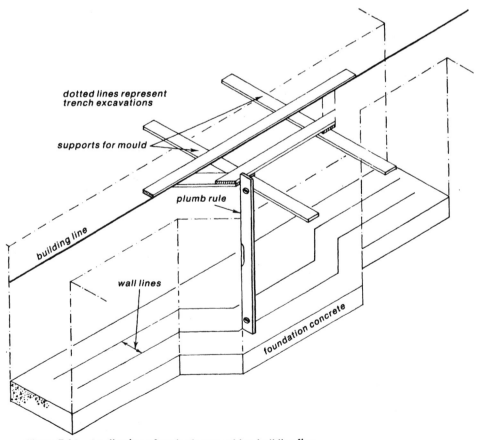

Figure 7.16   Application of squint bay mould to building line

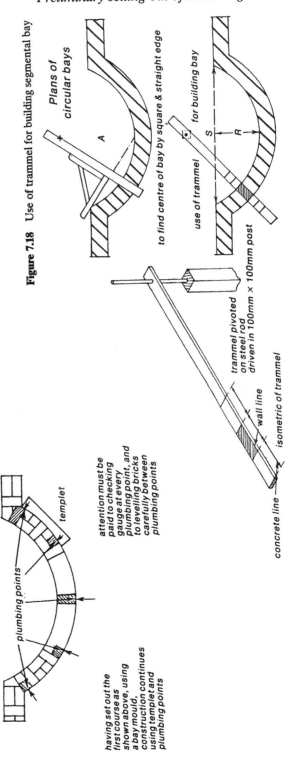

Figure 7.17   Use of segmental bay mould

Figure 7.18   Use of trammel for building segmental bay

use of templet between plumbing points

ranging line

application of circular bay mould to building line

templet

plumbing points

attention must be paid to checking gauge at every plumbing point, and to levelling bricks carefully between plumbing points

having set out the first courses as shown above, using a bay mould, construction continues using templet and plumbing points

Plans of circular bays

A

to find centre of bay by square & straight edge

use of trammel

for building bay

S

R

trammel pivoted on steel rod driven in 100mm × 100mm post

wall line

isometric of trammel

concrete line

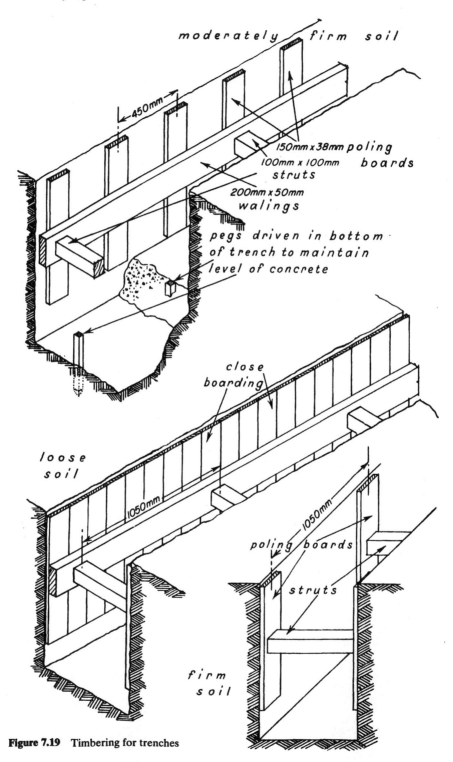

moderately firm soil

450mm

150mm x 38mm poling
boards

100mm x 100mm
struts

200mm x 50mm
walings

pegs driven in bottom
of trench to maintain
level of concrete

close
boarding

loose
soil

1050mm

poling boards

1050mm

struts

firm
soil

Figure 7.19   Timbering for trenches

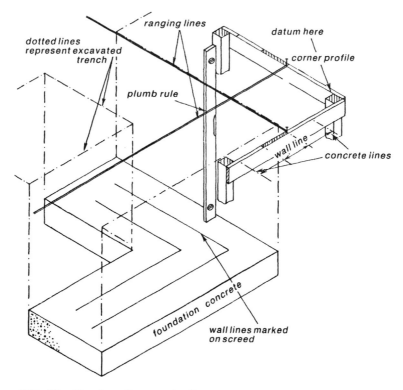

dotted lines
represent excavated
trench

ranging lines

datum here

corner profile

plumb rule

wall line

concrete lines

foundation concrete

wall lines marked
on screed

Figure 7.20   Plumbing down from ranging lines

3. The ranging lines are set to wall marks, and plumb lines are taken down on the main corners to the concrete foundation. To facilitate clear marking, a mortar 'screed' must be spread over the concrete to a thickness of approximately 3mm at the positions where the plumb lines are being taken. A plumb rule can normally be used to ascertain plumb lines, but if the trench is too deep to allow this to be done, a 'drop bob' should be employed (Fig. 7.20). Having completed these operations, proceed to erect corners with or without footings, as instructed, and check the neat work when the top course of the footings is reached.

4. The wall is filled in between corners, line and pins being used, and both corners and wall are continued to damp course level.

From this point, site concrete and damp-proof courses are placed. These are described in Chapter 10, but at this stage it is sufficient to know that the ground-floor level is approximately 225mm above damp-proof course level. When this point is reached in the building of the wall, a level must be transferred from the corner post of the profile which represents ground-floor level. This level will have been adhered to whilst constructing the wall, to ensure that the wall is horizontal at all points. The transferred level is maintained on the wall by fixing a short length of 50mm × 25mm batten, termed a 'datum peg' (Fig. 7.21). From this datum a 'storey rod' is used to maintain correct heights and, in order to be of assistance to the foreman

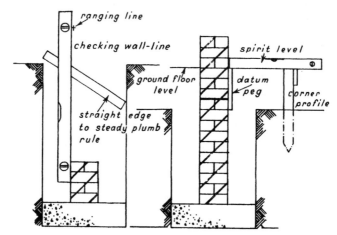

Figure 7.21 Transferring setting out lines
and datum to substructure

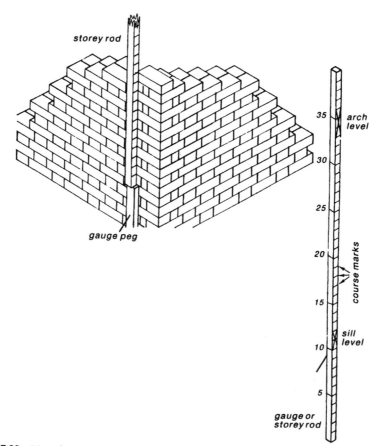

Figure 7.22 Use of storey rod

bricklayer, it should be marked with all those details appearing on the elevation of the wall, e.g. sills, arch and ventilator levels, etc. (Fig. 7.22).

Setting out of window openings.

Immediately a wall appears above ground level, preparation must be made in the bond arrangement for the insertion of window openings at a future height. At this stage, the proposed elevation of the building must be thoroughly studied and the work visualised so that every feature is known subconsciously, together with its correct position. In common brickwork which is to be covered, no special care is needed in the bond arrangement, provided all perpends are plumb, but where bricks form the face of a building, special attention must be paid to the bond arrangement. Whatever the type of bonding used, it must be arranged so that no straight joints occur when window openings are set out and the brick piers formed.

Figures 7.23 to 7.25 show a traditional design form; the dimensions of the windows and piers have been chosen to illustrate the procedure that should be followed in order to present a good appearance of the face brickwork.

Fig. 7.23 shows a complete elevation with the bond set out correctly. In this case a stretcher appeared immediately above ground level; if a header course had occurred it would have been arranged as at 'B' (Fig. 7.23). Note the quarter bond has been allowed for. The probable bonding below ground level is shown by dotted lines. It may be held that the correct bonding could have been arranged at footings level, but the bricks below ground level are often of a different type, and, in addition, the confined area of a trench makes work difficult, and precise setting out is not always possible. The positioning of broken bond will be obvious (see Chapter 4, page 53).

The bedding of artificial stone sills, which are built in as the building of the walls proceeds, is also shown. The sills must be bedded at their ends only, as if they are bedded throughout their length, the slightest settlement of the piers would cause them to fracture at the centre. The open joint can be made good when building is completed. When placing sills for alignment, always line or sight along the bottom edge; this is the 'eye-line.' Bed the end sills and line in all others to these. Where sills are not built in as the work proceeds, provision is made for their future insertion by building in 'sand courses,' as shown at 'A' (Fig. 7.23).

'Pinch rods' (Fig. 7.23) are always useful as they allow a check to be kept at all times on the correct sizes of window openings. These openings should never take the shape shown by the dotted lines, i.e. one reveal 'battered' and the other 'overhanging.' Pinch rods are not needed if window frames are built in.

Fig. 7.25 shows an exception to the usual method of setting out window openings. Examination of the two drawings will reveal that if the normal procedure is adopted (Fig. 7.24), four broken bonds occur, whereas if the problem is fully considered before the work is begun, a precise arrangement of bricks will require the use of one broken bond only.

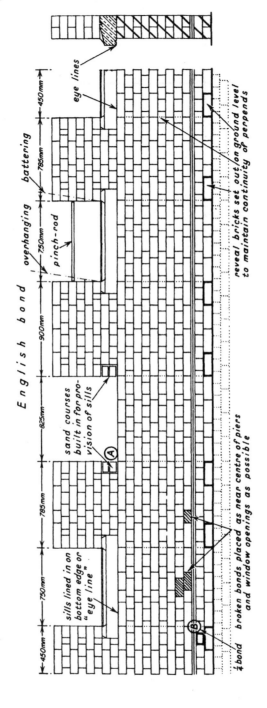

Figure 7.23 Setting out window openings (English bond)

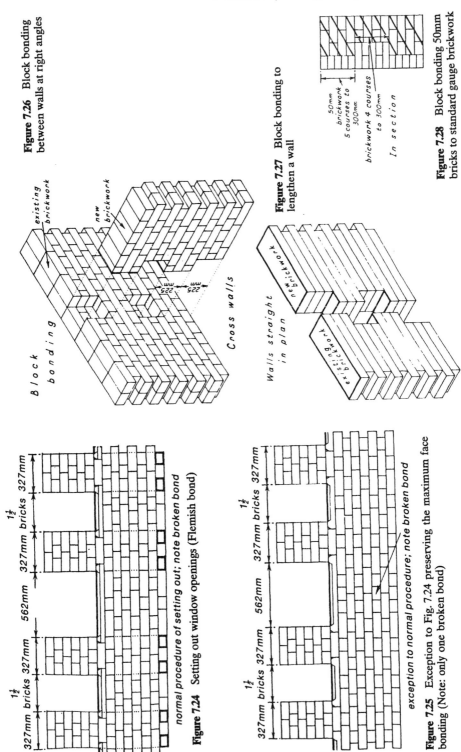

Figure 7.26 Block bonding between walls at right angles

Figure 7.27 Block bonding to lengthen a wall

Figure 7.28 Block bonding 50mm bricks to standard gauge brickwork

normal procedure of setting out; note broken bond

Figure 7.24 Setting out window openings (Flemish bond)

exception to normal procedure; note broken bond

Figure 7.25 Exception to Fig. 7.24 preserving the maximum face bonding (Note: only one broken bond)

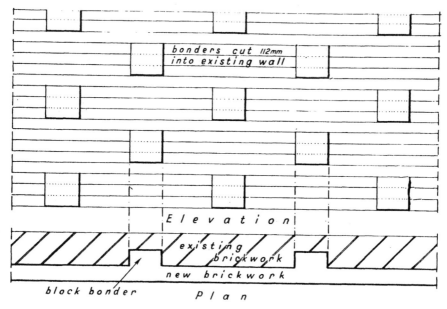

bonders cut 112mm
into existing wall

E l e v a t i o n

*existing
brickwork*

new brickwork

block bonder

P l a n

Figure 7.29 Block bonding to increase the thickness of a wall

Block bonding

A method adopted to bond a new wall into an existing building. Three types of work are shown:

1. Fig. 7.26. Block bonding a cross wall.
2. Fig. 7.27. Extending the length of a wall.
3. Fig. 7.28. Thickening an existing wall.

In Figs. 7.26 and 7.27 indented toothings could be cut and used for bonding purposes. In some circumstances this is a necessity, but block bonding should be used wherever possible, for two reasons:

1. A clear-cut hole and a more adequate bond are obtained.
2. The existing brickwork is not always of the same gauge as the new brickwork and attempts to bond on alternate courses are impractical.

In Fig. 7.29 block bonding is the most practical solution. Note that the blocks have been placed in a diagonal pattern, which gives greater surface bond than would be the case if the blocks were placed one above the other.

In the illustrations, block bondings have been cut away at every 225mm. This is the usual practice, but the bondings can be extended to 300mm without loss of stability.

In preparation for work of this type, the bricklayer cuts away the necessary brickwork with a cold chisel and club hammer. There are craftsmen who consider cutting away to be unskilled labour, but this is not the case, for the work does, in fact, require considerable skill. Quite often an apprentice employed on this work can be seen attempting to force the chisel, which has been entered into the face of the wall, by using the club hammer with both hands. No doubt expecting to see large pieces of brickwork fall away as a result, and disappointed when the brickwork splays and fractures in the wrong position or when the chisel disappears to its head and is difficult to extract. The art of cutting away requires that:

1. The chisel should never be forced.

2. A start should be made with a small hole, which should be cleared as cutting proceeds and which should be kept symmetrical as it is gradually enlarged.

3.The cutting edge of the chisel should be watched, and not its head.

Fig. 7.28 shows the block bonding of a wall in its thickness, when faced with 50mm facing bricks. Common bricks are not manufactured in 50mm sizes, and it would be too expensive to use 50mm facings throughout the thickness of the wall. A combination of both types of brick is therefore needed. A system of block bonding in the section of the wall is adopted to effect this, the blocks occurring at every 300mm, the wall being level at multiples of this height.

Toothings and indented toothings _____

It is occasionally found necessary to leave part of a building down and, in preparation for its future extension, toothings or indented toothings are formed as the work proceeds. In order to strengthen the connection between the walls when the extension is built one of the proprietary mesh reinforcements supplied in wall widths, are built in (Fig. 7.30).

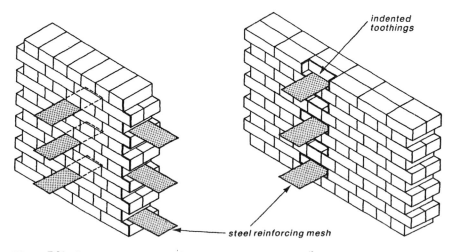

Figure 7.30 Proprietory mesh used to reinforce toothings or indents, and create a mechanical tie for blockwork where indents are not specified

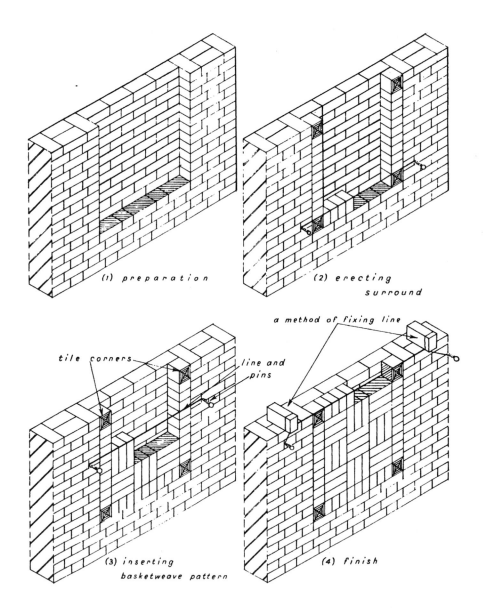

(1) preparation

(2) erecting surround

tile corners

line and pins

(3) inserting basketweave pattern

a method of fixing line

(4) finish

Figure 7.31　Building a flush panel

Decorative panels_____

Flush panels

Fig. 7.31 shows the necessary operations.

1. *Preparation.* The main brickwork is erected and provision is made for the insertion of the panel. The backing is built as the main work proceeds, as this gives adequate bond and assists when the panel is erected.

2. *Erecting the surround.* Tile corners are shown. The plumbing of the side must be watched and line and pins used for the base. Surrounds are not a necessity in this type of panel and have been added in this case as a decorative feature.

3. *Insertion of panel.* Whatever the type of bond adopted, it is always wise to make free use of line and pins, although a straight edge may be used for final adjustment.

4. *Finish.* Make sure that the top of the surround is level with the main brickwork. If it is low it gives an unsightly joint and if it is high, unnecessary cutting is involved.

Projecting panel (Fig. 7.32). Operations 1 and 3 are similar to those for flush panels. In erecting the surround, partly mitred brick corners have been adopted.

Note that the line and pins have been used on the bottom edge of the surround. This is an eye-line. With regard to the finish, a 25mm projection of the panel leaves a wide joint between backing and panel. A filling of fine concrete is the best practical method of dealing with this, as it saves excessive use of mortar and obviates the hollow gaps caused by attempting to insert too large a piece of brick.

Recessed panel (Fig. 7.32). Notice the backing in this case. A one-brick wall is illustrated and the backing is brick on edge, which can be block bonded in every 225mm. If the wall is thicker than one brick, a brick on edge backing is unnecessary.

Inserting the panel

In diagonal basket weave or herring-bone bonds the position of the bottom cuts is important. If the angle is lost it will be found necessary to shorten the length of a brick, which in turn will lead to a search for bricks of excessive length and eventually it may be found impossible to proceed further with the work, owing to the original loss of the angle. If a brick fails to fit into its correct position without cutting when the bottom cuts have been made, the work has been carried out incorrectly and the fault must be found before further work proceeds. When finishing, the eye-line must be noted.

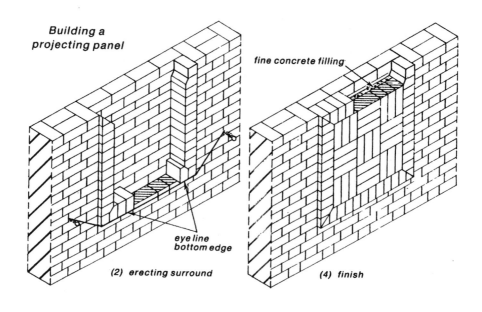

Building a projecting panel

fine concrete filling

eye line
bottom edge

(2) erecting surround

(4) finish

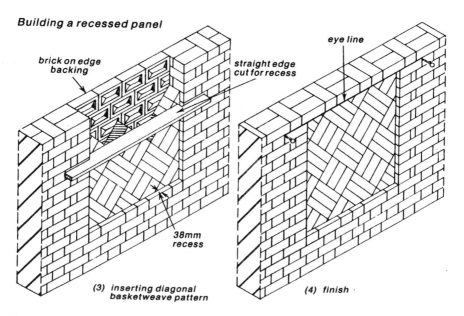

Building a recessed panel

brick on edge
backing

straight edge
cut for recess

eye line

38mm
recess

(3) inserting diagonal
basketweave pattern

(4) finish

Figure 7.32

Cutting of gable ends _____

(Figure 7.33). The illustration is self-explanatory. The correct position of the gallows must be ascertained, both for angle of line and for plumb above the main brickwork. If the latter is correct, the remaining brickwork need not be plumbed. Note the wind-filling between the rafters; this work is usually carried out by the apprentice and is excellent practice in the use of the trowel. It may become monotonous and the apprentice may be tempted to lose interest, but it affords good training and the study of roof construction is an absorbing one.

Cutting brickwork to a lean-to

(Figure 7.34). The brickwork must be racked back to obtain the correct cut and a fixing for the line and pins. Note the brick corbels supporting the 'wall plates.' These are usually 100mm × 50mm timbers, and they are used to distribute the weight of the roof and to obtain a firm fixing. The bricklayer has to bed these wall plates and this must be done solidly and level, using a 3 m to 4 m straight edge of timber or hollow section aluminium and spirit level.

Procedure for raking cutting _____

1 Rack back and tooth out below the sloping line and pins as shown in Fig. 7.35.
2 Set the sliding bevel at the necessary angle of the raking cutting.
3 Measure the top edge of the cut brick required.
4 Deduct 10mm from one cross joint.
5 Mark in pencil the remaining measurement on the brick to be cut, back and front.
6 Draw in pencil the angle of sloping (raking) cut, using the sliding bevel, also back and front, taking care to match the slope on the face.
7 Cut the brick with hammer and bolster.
8 Trim the cut face with scutch or comb hammer, slightly 'undercutting' if possible to avoid any projections which would prevent the coping or capping from being bedded properly.
9 Carefully bed the cut-brick to line and level.

Alternative method, if sliding bevel is not available _____

1 Rack back and tooth out as before.
2 Temporarily bed a brick, carefully, to line and level, as shown in Fig. 7.35 at bottom left.
3 Apply two pencil marks on the brick face to indicate the angle of cut.
4 Remove the brick from its temporary bed, and cut it with hammer and bolster; trim with scutch; and permanently bed the cut brick.

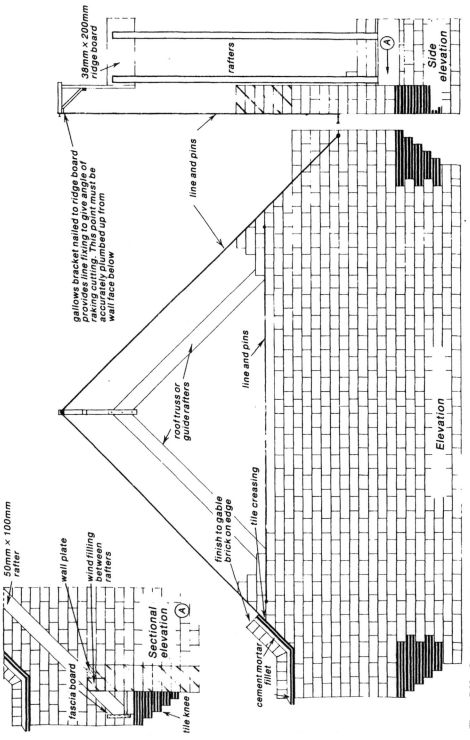

38mm × 200mm ridge board

gallows bracket nailed to ridge board provides line fixing to give angle of raking cutting. This point must be accurately plumbed up from wall face below

rafters

line and pins

A

Side elevation

roof truss or guide rafters

line and pins

finish to gable brick on edge

tile creasing

Elevation

50mm × 100mm rafter

wall plate

wind filling between rafters

fascia board

Sectional elevation   A

tile knee

cement mortar fillet

Figure 7.33   Cutting up a gable end

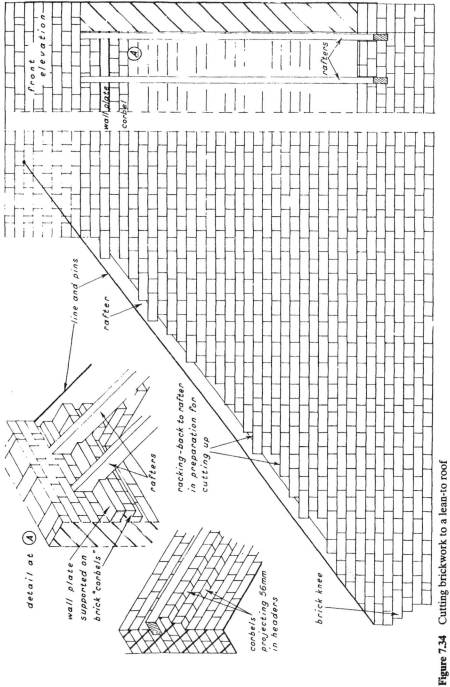

Figure 7.34 Cutting brickwork to a lean-to roof

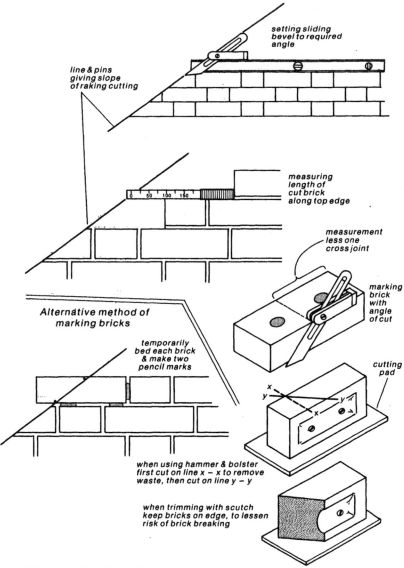

setting sliding
bevel to required
angle

line & pins
giving slope
of raking cutting

measuring
length of
cut brick
along top edge

measurement
less one
cross joint

marking
brick
with
angle
of cut

Alternative method of
marking bricks

temporarily
bed each brick
& make two
pencil marks

cutting
pad

x
y
y
x

when using hammer & bolster
first cut on line x – x to remove
waste, then cut on line y – y

when trimming with scutch
keep bricks on edge, to lessen
risk of brick breaking

Figure 7.35   Procedure for raking cutting

If perforated wirecut facing bricks or engineering bricks are being used, these will have to be cut on a masonry bench saw. Mark five or six at a time, indicating clearly the 'waste' part of each brick to be cut. Number each one for easy identification, and send them for cutting on the site masonry bench saw.

Safety

Warning. It is an offence under the Health and Safety at Work Act to use a masonry bench saw if you are under the age of eighteen. Even aged eighteen and over, the Act says you must have been instructed in its safe use. So follow these simple but vitally important rules:

Wear eye and ear protection

Ensure that you are wearing no loose clothing

The blade must be secured and fixed by a qualified person

Check that the blade guard is in place

Clamp the brick to be cut firmly on the trolley, and keep your fingers well away from the blade

Make sure that the water tank is filled and the pump working

Do not force the blade as it cuts through the brick.

Brick corbels _____

(Figures 7.36, 7.37 and 7.38). The building of oversailing courses forming corbels—in 56mm oversailers for the laying and in 28mm for bond arrangement—is not an easy operation to perform. In the case of the 56mm corbels, headers should be used where-ever possible; if the use of stretchers cannot be avoided, they should be the last bricks laid on that course and should be well bedded, a cross-joint being placed throughout the length of the brick. In the case of 28mm corbels, the greatest possible lap must be

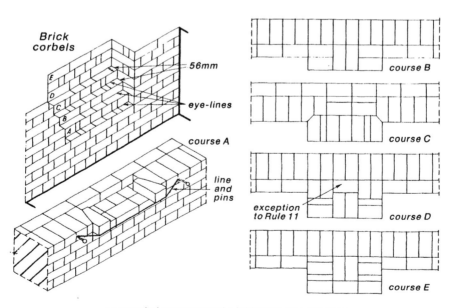

each corbel course must be laid to the line, but fixed to the bottom arris, as this forms the 'sight-line'

Figure 7.36 Setting corbel courses to a line

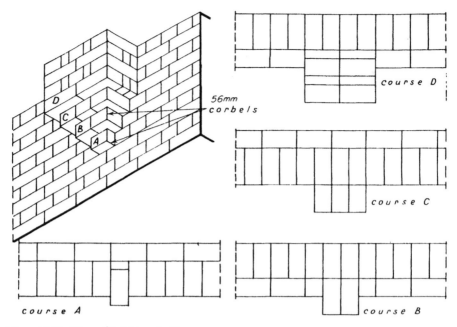

Figure 7.37   56mm (¼ brick) corbelling

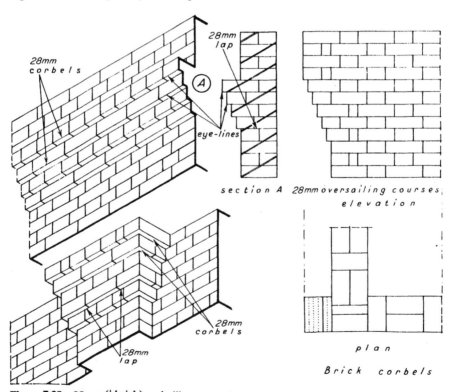

Figure 7.38   28mm (⅛ brick) corbelling

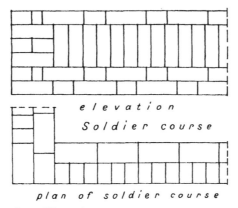

Figure 7.39 Constructing soldier string course

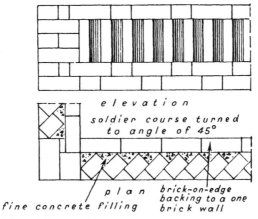

Figure 7.40 Soldier bricks turned to an angle

maintained; if special attention is given to this point, and common sense applied, no difficulty should be experienced in carrying out this work.

Both internal and external straight joints are sometimes unavoidable, but they should be reduced to a minimum and the specified bond should be adhered to, if possible. The final shape of the brickwork must be considered; this will continue to a greater height than the corbel and must be practical and not involve unnecessary cutting.

When a corbel is being built it must be kept well tailed down to prevent overturning, that is, the back bricks must be laid first in order to bond the previous corbel before attempting to lay the next oversailing course.

'Soldier' course

(Figures 7.39 and 7.40). Consists of bricks laid on end continuously. Care should be taken to avoid laying the course at an angle and a small boat-level should be used for checking the work see Fig. 7.41.

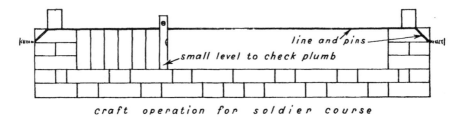

craft   operation   for   soldier   course

Figure 7.41 - Checking soldier course bricks

Dog toothing

(Figure 7.42). The correct angle must be maintained throughout and alignment must be kept. 'A' shows the treatment in the centre of a wall, the work having been carried out from both ends.

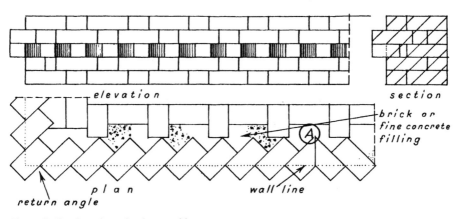

Figure 7.42   Creation of a dog toothing course

Dentil course

(Figure 7.43). No explanation is needed, except that the eye or sight lines must be maintained.

String or band courses are a decorative feature and are quite separate from normal bonding arrangements. Straight joints between main brickwork and the string course must be avoided wherever possible.

Many other designs are possible for string courses by the introduction of basket weave and herring-bone bonds, tiles or bullnose and cant special bricks set as soldiers. All these are laid in the manner already described, eye-lines must be maintained and bonding knowledge applied.

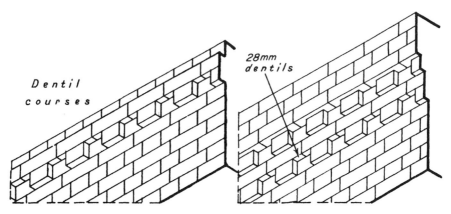

Figure 7.43 Single and double dentil courses

Tumbling-in

(Figure 7.44). This is a method of reducing the size of a pier or external chimney breast. It is decorative in character and acts as a weathering as shown. Figure 7.45 shows the preparation, building, and application of buttress tumbling-in for an external chimney. 50mm × 25 mm batten is a temporary fixing for line. The over-hang at 'A' forms a 'drip,' thus preventing rain from running down the tumbling directly on to the wall face, and it avoids the ugly joint which would occur if no over-hang were introduced. The illustration shows a diaper pattern introduced into a chimney breast as a decorative feature.

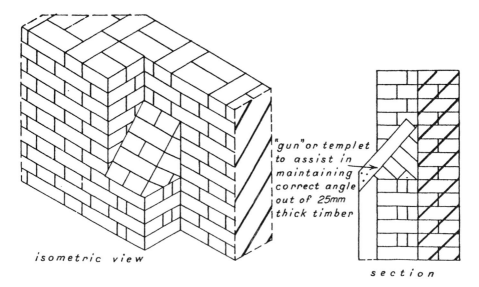

Figure 7.44 Small section of tumbling in using a gun templet

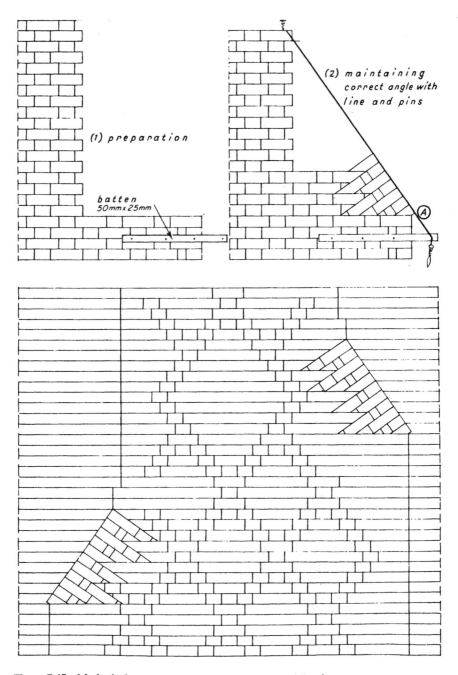

(1) preparation

batten
50mm x 25mm

(2) maintaining
correct angle with
line and pins

Figure 7.45 Method of constructing longer sections of tumbling-in

Brick ramp

(Figure 7.46). The circular ramp shows a brick-on-edge finish, but a brick laid flat may be used, according to the architect's design.

Fig. 7.47 shows the method of maintaining the stability of the circular ramp by building in 200mm long stainless steel fish-tail cramps. Brick-on-edge finish, whether circular or straight, can be strengthened in this manner.

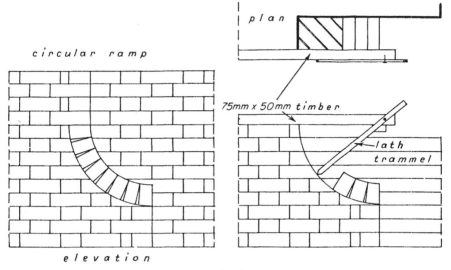

Figure 7.46 Use of trammel to construct a circular ramp

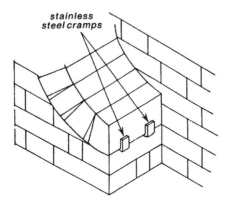

Figure 7.47 Stainless steel cramps securing first ramp bricks

8
Bridging openings

This area of bricklayer's work is about permanently holding up walling over the top of doors and windows. Early civilisations tried various ways of supporting brick and stone masonry across openings in their buildings as shown in Figs. 8.1 and 8.2.

Pyramid-building Egyptians and ancient Greeks used lots of columns and stone lintels, but it was the later Roman builders who developed the idea of forming arches from separate blocks of stone, bricks or tiles mortared together.

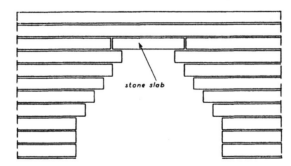

Figure 8.1 Spanning an opening by corbelling

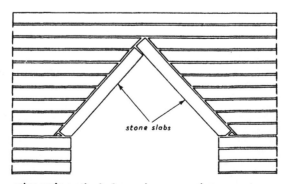

Figure 8.2 Alternative early method of spanning an opening

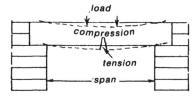

Figure 8.3 Typical stresses in a lintel

Definitions

Lintels

These are straight beams of concrete or steel which are made strong enough to support the dead load of bricks or blocks above. Because lintels are straight, they have a slight tendency to bend or deflect when bricks or blocks are bedded on top. This loading results in tensile stress in the lower part of a lintel, plus compressive stress in the upper part, as indicated in Fig. 8.3. Steel rods are cast into concrete lintels to absorb this tension, so that the lintel does not crack on the underside when loaded. Concrete in the upper part of the lintel is very good at resisting compressive stress.

Sheet steel can be pressed into shape in a factory to make lintels that are much lighter to lift than concrete, and which can withstand both the tensile and compressive stresses, see Fig. 9.11(a).

Brick arches

Arch construction is a more decorative means of spanning openings. True arches are made to curve upwards when looked at in elevation, so that they are always in a state of compression, wedged between abutments. As bending under load cannot take place, there will be no tensile stress in the arch.

All arches are formed upon a temporary support called an arch 'centre' which must not be removed until the jointing mortar has been allowed to harden for at least a week. Only when the arch centre has been removed does the arch become self supporting.

Lintels

Reinforced concrete lintels

Properly bonded brickwork is able to 'bridge' itself across openings in a series of offsets as shown in Fig. 8.4. This is called 'natural bracketing' and means that in theory, only that area of brickwork within the 60 degrees triangular shape in Fig. 8.4 needs supporting by a lintel. The greater the span of an opening, the larger this triangular piece will be.

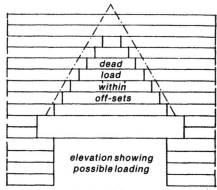

Figure 8.4   Natural bracketing of bonded brickwork above a lintel

Lintel bearings

A solid seating or bearing is required at each end of a lintel, to support the concentrated pressure of one half of the total load resting upon the lintel. This bearing at each side should be a minimum of 150mm for lintel spans up to 1800mm, with 225mm minimum up to 3m span. Thereafter, a Structural Engineer will be required to calculate the safe bearing area needed for larger span lintels. BS5628:1985 Part 3 'Workmanship' recommends that this bearing level should coincide with a *whole* block rather than a half or cut block, for greater stability where lintels are set in block walling. See Fig. 8.5.

Bedding lintels

Lintels must be set upon a wet mortar bed joint at each side, so as to spread the load evenly over the whole bearing surface. When tapping down into position the spirit level should be held against the underside of the lintel in case the top edge is not parallel with the soffit.

Types of reinforced concrete lintel

Concrete lintels can be made available in any one of five ways, depending upon the following site considerations.
- How large lintels need to be.
- How much each will weigh (mass in kg).
- How they are to be moved around site and to upper floors.
- Are they to be lifted into place by hand or mechanically?
- Will they be seen in the finished work, or plastered?

Precast lintels

These are lintels that are made in a mould at ground level, and when the concrete has hardened for at least a week, are lifted and bedded in place.

Fig. 8.6 shows a typical site casting platform, set perfectly level both ways. Bricks or blocks are used as spacers, in order to obtain lintels of exactly the right width.

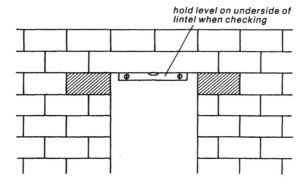

Note: This means setting out the first course to ensure
that this happens

Figure 8.5 Whole block under lintel bearings to spread load

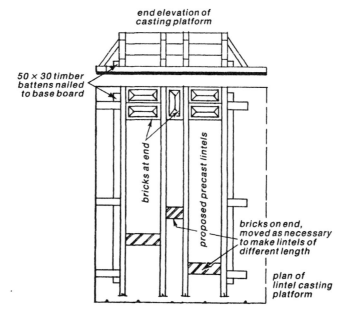

Figure 8.6 Method of casting lintels on site

Inside surfaces of these lintel 'boxes' should be lightly coated with mould oil to make dismantling or 'striking' easier. When filling the mould a 25–30mm layer of wet concrete should be spread over the bottom first before placing reinforcing steel bars. The remainder of the concrete should be added and thoroughly compacted by manual or mechanical vibration.

The upper surfaces should be trowelled smooth and marked clearly as the 'Top' surface, so that the steel bars will be towards the soffit when permanently bedded.

Figure 8.7 Form work for a cast-in-situ lintel

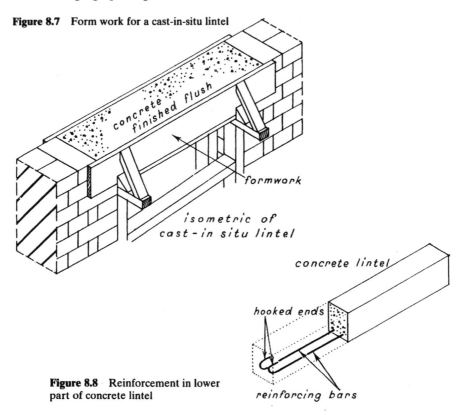

Figure 8.7 isometric of cast-in situ lintel

Figure 8.8 Reinforcement in lower part of concrete lintel

Factory made reinforced concrete lintels

For larger requirements precast r.c. lintels can be ordered from a specialist supplier to suit the range of openings in a building. Care should be taken to see that a 'fair-face' is specified where lintels will be permanently exposed, and that the 'top' surface of all lintels delivered to site is clearly marked.

Prestressed concrete plank lintels

These are a special type of r.c. lintel which can only be made in a factory. The high tensile steel 'strands' are stretched by hydraulic jacks *before* the concrete is compacted in the lintel mould. When the concrete has thoroughly hardened, the stretched steel reinforcement is carefully released and keeps the whole lintel section in a permanent state of compression. These lintels are known as 'plank' lintels because they are only 65mm deep for all spans up to 1800mm.

Cast in-situ *concrete lintels*

This is where a lintel would be much too heavy to lift by hand if precast at ground level.

Instead, a timber box or mould is constructed exactly where the lintel is required. This formwork, Fig. 8.7, is made strong enough to withstand the pressures of filling with wet concrete, and thoroughly compacting it. The concrete is poured around steel reinforcing bars placed approximately 25mm up from the soffit, see Fig. 8.8.

Reinforced brick lintel

The bricklayer calls this a 'soldier arch', as the bricks are placed on end resembling a file of soldiers. Before setting the lintel, temporary supports are placed (see Fig. 8.59). Several methods of reinforcing brick lintels are shown (see Fig. 8.9); at 'A' where a 225mm soffit occurs, steel bar reinforcement is placed throughout the length of the lintel, with stainless steel ties positioned at intervals of approximately 225mm; at 'B' the brick lintel is erected with wire reinforcement built in and left projecting, to allow the concrete lintel to be cast *in situ* around it, thus giving adequate stability. This type of lintel, if erected truly horizontal and plumb, causes an optical illusion to occur at a distance of 3 metres or more from it; it appears to sag. This appearance can be overcome by giving the lintel a slight camber or by giving a curvature to the soffit along its length, together with a slight skew-back. The adjustment must be made carefully and with the consent of the architect.

Classification

Arches

The names of the various parts of an arch are shown in Fig. 8.10. Arches are named or classified in the following ways.
   1. The method of cutting or fixing, e.g. rough ringed, axed, gauged;
   2. their shape, e.g. Gothic, camber or flat, semi-circular, segmental;
   3. the number of centres from which they are set out.
   These classifications can be used in combination, e.g. axed segmental, gauged three-centred semi-elliptical.

Rough ringed arch.   This type of arch is the simplest to construct. It is formed of uncut bricks and its shape is controlled by the type of turning piece or centre adopted. The joints are wedge-shaped and thus play an important part in the stability of the arch. In the axed or gauged arch the reverse applies—bricks are wedge-shaped and the joints are parallel. The arch is constructed of a number of half-brick rings, hence the name ringed arch, the number of half-bricks varying according to the size of the opening. This method is adopted to reduce the size of the mortar joint at the extrados of each separate ring (Fig. 8.11). In the illustration, the use of stretchers to turn the arch would make the width of the joint at the extrados excessive.

   Over large spans, and where the arch is several rings in depth, 'lacing courses' (three or four courses of axed work) can be inserted at intervals throughout the depth and thickness of the arch as a decorative feature (Fig. 8.13). Figure 8.12 shows a $1\frac{1}{2}$-brick arch made up of two rings, with the introduction of lacing courses at skewback and key. It will be noted that the bottom ring is one brick in depth and bonded; this is possible where the arch is of a wide span and of easy curvature.

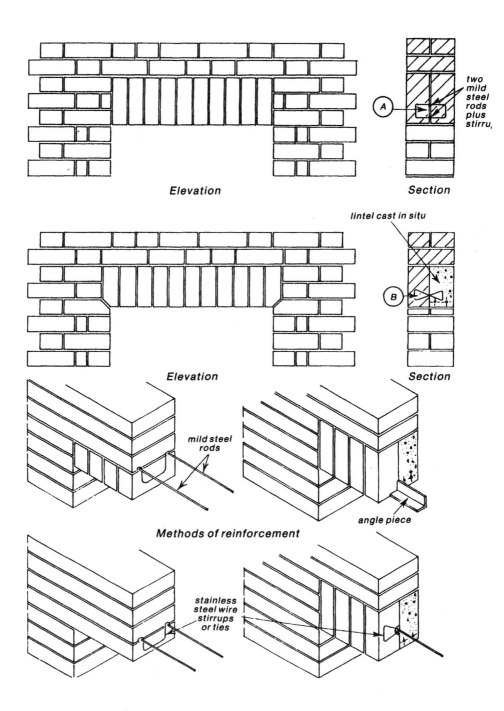

Figure 8.9   Soldier arches and reinforced lintels

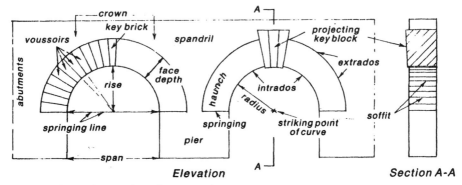

Figure 8.10 Craft terms in arch construction

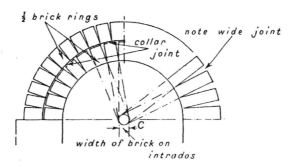

Figure 8.11 The reason for using two header rings and not stretchers for rough arches

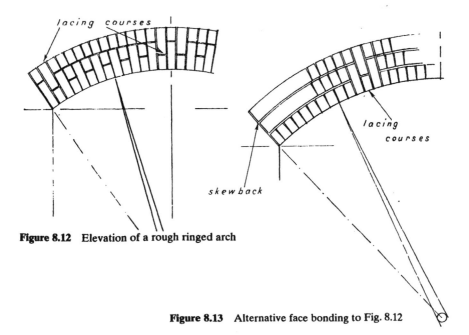

Figure 8.12 Elevation of a rough ringed arch

Figure 8.13 Alternative face bonding to Fig. 8.12

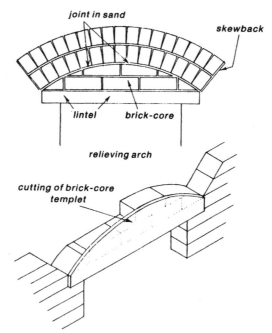

Figure 8.14   Formation of relieving or discharging arch

In the days when walls were built of flints, horizontal bands of brickwork, three or four courses deep, were introduced at various intervals. These were called 'lacing' courses, and used in this way they helped to strengthen the wall structure.

Relieving or discharging arch.

This is a form of ringed arch, built over a lintel, and turned on a shaped brick core (Fig. 8.14). It relieves the lintel of any point load and discharges the weight to the abutments. The joint between the brick core and the soffit of the arch must be of sand, as this ensures the proper discharging of loads. This sand joint is pointed on the face only. The shape of the brick core is obtained by the use of a templet.

Axed arch.

This type of arch is formed by bricks cut to wedge shape from the ordinary facing brick. It can be of the same colour and texture as the wall facings or of some contrasting colour and probably of finer texture. The tools used for cutting and trimming are the hammer, bolster, and scutch, with a carborundum block.

The thickness of the mortar joint varies from 4mm up to 10mm.

The operations of cutting and setting will be explained later in this chapter.

Brick arches—gauged work

Earlier editions of this book contained a whole chapter describing the bricklayers' craft operation of cutting and rubbing gauged arch brickwork. This work requires specially made clay bricks called 'red rubbers'.

These bricks contained approximately 30% fine sand mixed with the clay, and were carefully fired so that the same even orange-red colour and texture was present throughout the body of each brick.

The high content of sand allowed these bricks to be hand sawn and rubbed, on a block of York stone, to permit joints of 1mm thickness. Gauged work demanded the very highest skills from the bricklayer in setting out, cutting, rubbing and bedding these bricks, which were 'white-line' jointed, using a putty made only from freshly slaked lime and water.

Much thought was given before deciding to omit this chapter (which fully described this highly skilled aspect of the bricklayer's work) from this revised edition. The primary reason for leaving it out is that these red rubber bricks are no longer available; another reason is that modern methods of cutting voussoirs on masonry bench saws have displaced the labour-intensive traditional method of cutting and rubbing by hand.

Masonry bench saws can cut plain red facing bricks to shape without the high sand content, and still obtain the fine 'white-line' joints that distinguish the appearance of gauged work from other face brickwork.

Types of arch _____

This section will deal with the shape of arches and their geometrical setting out. These arches will all be shown straight in plan. Various simple geometrical constructions are illustrated and explained to assist in the understanding of arch setting out and correct construction on site, see Figs. 8.15 and 8.16.

1. *To bisect a given line AB*. With the compasses set at a radius greater than half the length of the line, and used successively at points A and B, describe the intersecting arcs AC, AC' and BC, BC'. Then CC' is the perpendicular bisector.

2. *To bisect the angle ABC*. Set the compasses at any radius and from point B describe arc DD'. With centres at D, D' alternately, and with the same radius or a radius greater than a half DD', describe the intersecting arcs E. Then BE is the bisector.

3. *To erect a perpendicular on line AB from point A*. From any centre C and with the distance CA as radius, describe a circle cutting AB at D. Draw the line DCE. Then the line AE is the perpendicular.

4. *To divide a given line into a given number of equal parts*. From one end of the line AB, draw a straight line at any angle. From point A on this line mark off the given number of equal lengths, say 5. Join B5. Through the other points draw parallels to the line B5. Then the line has been divided into five equal parts.

5. *To draw a tangent through a given point C on the circumference of a circle*. Draw the line AB from the centre of circle to cut the circumference at C. the given point. Make CB equal to AC and bisect AB by drawing arcs

D and E as in Method 1. Then line DE is at a tangent to the circle.

Note that the intrados and extrados of an arch are made up of a series of tangents through its voussoirs. A line taken through the centre of a voussoir from the striking point is at right angles to this tangent (Fig. 8.15).

6. *To draw a circle to pass through three given points A, B, and C.* Join AB, BC. Bisect the lines or chords AB, BC. Then the centre of the required circle, point D, is at the point where the perpendicular bisectors intersect (Fig. 8.15).

7. Fig. 8.16 illustrates circles in external and internal contact. They touch one another and are said to be at a tangent. Notice that their centres are on the same line AB, which passes through the point of contact and that a tangent drawn at C is common to both circles. Line AB is called the 'common normal.'

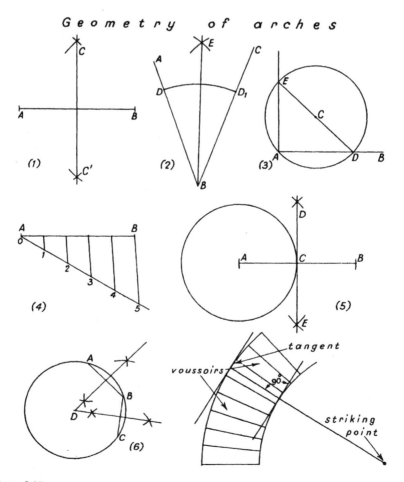

Figure 8.15

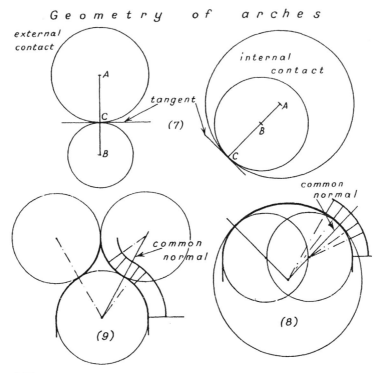

Figure 8.16

8. An application of 7 is that circles in internal contact form the curve of a three-centred or semi-elliptical arch. Note that a three-centred arch is not truly elliptical; it follows a similar shape and is therefore called elliptical, but if a true ellipse were formed, each brick would need a separate templet. By forming an approximate ellipse of three or five centres only, two or three templets respectively are needed. This will be further explained in a later section.

9. A further application of 7, showing circles in external contact forming an ogee curve.

Segmental arch

Figure 8.17 shows the geometrical setting out of a segmental arch. The rise to any segmental arch is normally one-sixth of the span. First draw the span, assume it to be 900mm, and set up a perpendicular bisector. Mark off the rise, 150mm. Join AB and bisect. Where this bisector intersects the centre line, point C, is the striking point of the required arch. A face depth of 225mm is shown.

Figures 8.18 and 8.19 show the elevation of ringed and axed arches.

To ascertain the number of voussoirs and the position of the joint lines in the axed arch, set the dividers at 75mm, place the points equidistant on each side of the centre line, thus forming the key brick and step round the extrados. If the last step fails to connect with the springing point at the first

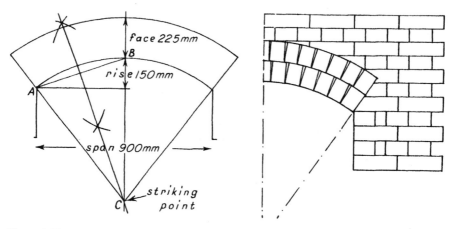

Figure 8.17 Drawing of segmental arch with rise of one sixth of span

Figure 8.18 Two ring rough segmental arch

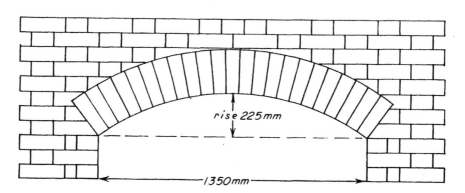

Figure 8.19 Axed segmental arch

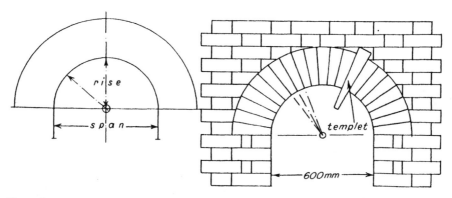

Figure 8.20 Construction of a semi-circular arch

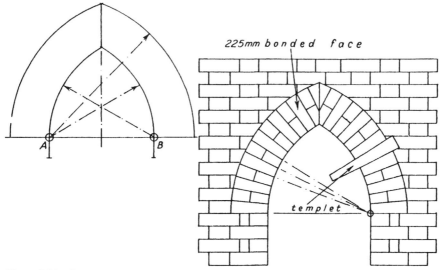

Figure 8.21 Construction of an equilateral Gothic arch

attempt, make a further attempt by slightly closing the dividers. The equivalent size, 75mm, is accepted in general drawing practice for both axed and gauged arches. For the full-size setting out, however, in preparation for the cutting of an axed arch, the voussoir size is taken from the type of brick being used.

There is no need in practice to draw or set out a ringed arch, but for drawing to scale, as a means of illustration, draw the intrados and extrados, step out the brick size on the intrados and obtain the wedge-shaped joints by describing a circle equal to the size of brick, at centre C (See Fig. 8.11, page 131).

Semi-circular arch

Figure 8.20 is self-explanatory. Note the joints radiating from the centre or striking point. One face templet is needed for the cutting of this arch.

Gothic arch

All arches under this heading are of the pointed type, and each has its particular name. Observe that no key brick occurs.

Equilateral (Fig. 8.21). With point A as centre and AB as radius, describe an arc from springing line to the centre line. The intrados and the extrados are concentric (i.e. have the same centre). Repeat movements on the opposite springing, with point B as centre. This is a two-centred arch. Complete the setting out of the voussoirs as in the segmental arch, except that there is no key brick. One face templet is needed.

Drop Gothic (Fig. 8.22). Draw the springing line and assume a span of 600mm. Set up the centre line and mark on this the rise; assume this to be

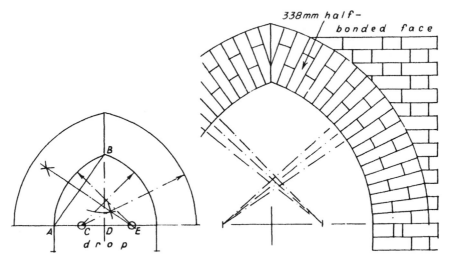

Figure 8.22 Construction of a drop gothic arch

400mm. Draw the chord AB and bisect. The point where the bisector cuts the springing is the striking point of the arc required. To obtain the opposite striking point, make DC equal DE. The striking points are *inside* the springing points.

Lancet (Fig. 8.23). Draw the springing line and extend it beyond the reveal lines. Mark the span 600mm and the rise 675mm, and proceed as in the drop arch. The striking points fall *outside* the springing points of the arch.

The position of the striking points denotes the type of arch. Thus, for 'drop' or 'lancet' arches the striking point is inside or outside the reveal lines respectively, while for the 'equilateral' arch it is constant, that is, on the springing points.

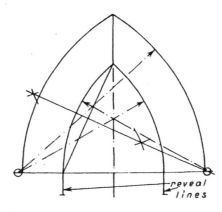

Figure 8.23 Construction of a lancet arch

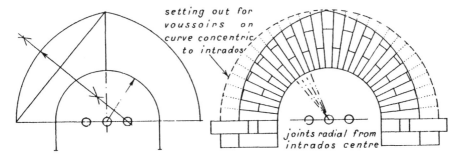

Figure 8.24   Florentine or semi-gothic arch

Semi-Gothic   (Fig. 8.24). This is an arch with a semi-circular intrados and a Gothic extrados. The depth of the face of the arch at the springing is 225mm and at the crown 338mm. To obtain the Gothic extrados, follow the method used for the drop or lancet arches. The setting out of the voussoirs on the extrados is taken on a curve which is concentric to the intrados; this regulates the size of the voussoirs. Note the key brick.

Venetian Gothic   (Fig. 8.25). The intrados and extrados are Gothic in shape, but have a different radius, given by a 225mm face depth at the springing and a 338mm face depth at the crown. The voussoir sizes are obtained by the method used for the semi-Gothic—a curve concentric to the intrados. Fig. 8.25 illustrates the cross-joints of the arch concentric to the intrados, and Fig. 8.26 shows them symmetrical to both intrados and extrados. The method adopted depends on the architect's instructions. To find the centres for the cross-joints in Fig. 8.26, divide the face of the arch at the springing into three equal parts and follow the same procedure at the crown. Assume the parts to be complete arcs, construct the chord lines and bisect, following the methods shown for striking the intrados and extrados arcs.

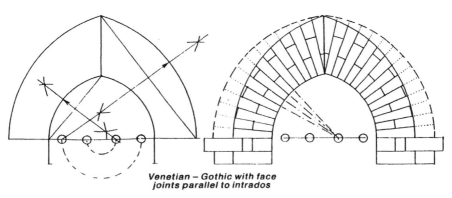

Figure 8.25

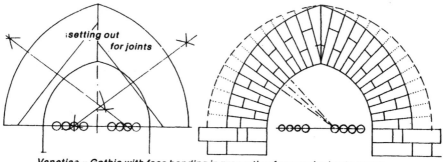

Venetian – Gothic with face bonding in proportion from springing to crown

Figure 8.26

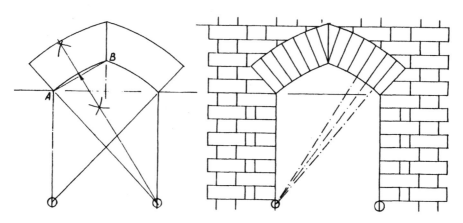

Figure 8.27   Segmental gothic arch

Segmental Gothic (Fig. 8.27). A Gothic arch with a skewback, which probably derives its name from this fact. It is formed by two segments meeting and forming a pointed arch. Set out the springing line and the rise. Join AB and bisect; where the bisector intersects the reveal line the striking point of one half of the arch occurs. Mark a similar striking point on the opposite reveal to complete the setting out.

Elliptical or Tudor Gothic   (Fig. 8.28). This is a four-centred arch, the composite curve being constructed by the application of circles in internal contact. Several methods can be adopted for the geometrical setting out of this arch, but the method illustrated is universal, as it permits of the construction of the arch to a given rise and a given span. Two face templets are needed for its cutting. The setting out is shown in its various stages:

1. Assume the rise to be 450mm, and the span to be 900mm. Draw the springing line and set out the span. Erect the perpendicular bisector

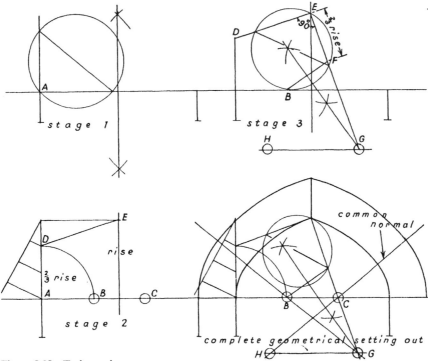

Figure 8.28   Tudor arch

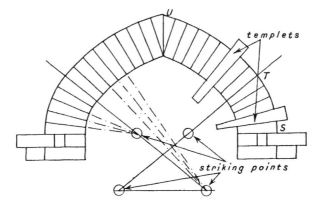

Figure 8.29   Elevation of Tudor arch

between springing points and set up a perpendicular at A.

2. Mark off the rise from the centre point of span to E, and set out two-thirds on the perpendicular at A. With the point of the compasses at A and with a radius of two-thirds the rise, describe a quadrant to intersect the springing line at B. This is the first striking point. Ascertain point C equidistant from centre line at B, to make the second striking point. Join DE.

· 3. Draw a line at 90° to DE at point E and on this mark off two-thirds of the rise EF. Join F and the first striking point B, and bisect. Where this bisector meets the line from E produced, the third striking point G occurs. Ascertain point H equidistant from centre line as G, to make the fourth striking point. Note the common normals through striking points GB, HC. In all arches where different curves are employed, make the common normal a joint line and set out voussoir sizes on the respective curves on each side of this. Thus in Fig. 8.29, X bricks between S and T, and similarly between T and U. Fig. 8.30 shows an arch with a face depth of 338mm bonded quarter-bond and Fig. 8.29 an arch with a 225mm face and the two templets required for the cutting.

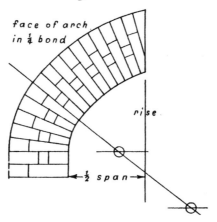

Figure 8.30 Half elevation of Tudor arch, bonded on face

Semi-elliptical or three-centred arch

(Figure 8.31). Another example of circles in internal contact. As in the case of the Tudor arch, several methods are possible in the geometrical setting out, but the method shown is general. Assume the rise to be 300mm and the span to be 900mm.

1. Set out the springing line and the rise.

2. Join AB. With the point of the compasses at C and with a radius equal to half the span, describe a quadrant to intersect the centre line at D. With the point of the compasses at B and with a radius equal to BD, describe an arc to cut AB at E.

3. Bisect the line AE, and where this bisector intersects springing line and centre line the first and second striking points occur respectively. The third striking point G is on the springing line equidistant from the centre line, as F.

The figure is self-explanatory.

As previously explained, the three-centred and five-centred arches are approximate ellipses. The more centres used in the geometrical construction, the truer the ellipses become. For an arch span of up to and including 1350mm, three centres give a pleasing curve. Beyond this size and up to, say, 3 m—the span limit which a bricklayer meets in general practice—five centres should be used.

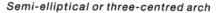

Semi-elliptical or three-centred arch

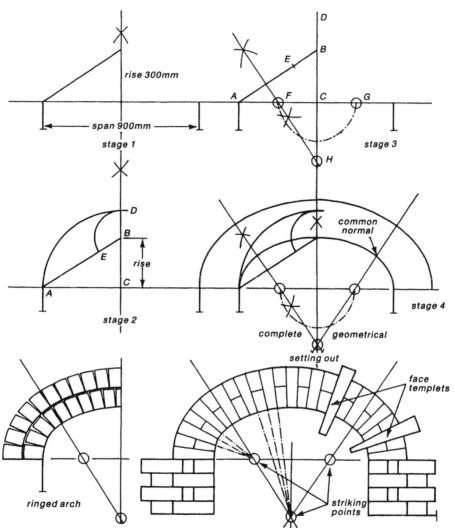

Figure 8.31 Semi-elliptical or three-centred arch

Camber or flat arch

(Figure 8.32). The name of this arch is explained by the fact that the arch is intended to be perfectly horizontal, but in order to prevent the optical illusion of sagging, it is given a slight camber or rise. At one time, the arch was given a very acute skewback, but it was found to fracture across the top points of the skewback.

To set out:

1. Draw the springing line, and parallel to this and 300mm above it, draw the setting out line.

2. Erect a perpendicular at the springing point to intersect the setting out

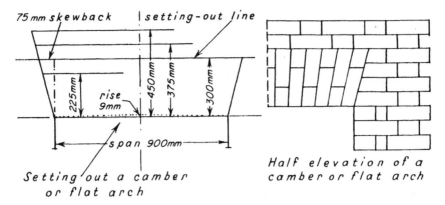

Figure 8.32   Camber or flat arch

line. From this intersection, mark off the skewback, allowing 25mm of skewback to every 300mm of span. Thus, for a 900mm span, a 75mm allowance is given; for a 1350mm span, 112.5mm; and so on. This inclination is constant for any depth of arch face (Fig. 8.32).

The amount of camber given to the soffit is 3mm to every 300mm of span, thus for a 900mm span the rise is 9mm, and for a 1350mm span, 14mm.

In the setting out of the voussoirs in a camber arch, some architects require a stretcher to occur on the soffit at both the key and skewback (Fig. 8.32). To effect this, the number of voussoirs must be a multiple of four, plus one. This procedure is not always adopted, however, as it is sometimes not considered to be economical or practical. Wherever possible, place a stretcher at the key with either a stretcher or header at the skewback, as the case may be.

Moulded segmental arch

(Figure 8.33). The setting out for the segment is the same as the construction previously explained, but this arch differs from the ordinary segmental arch at the springing point. Where the mouldings on the reveal and soffit of arch intersect, a mitre occurs and the skewback is projected from this, as illustrated. Two methods of obtaining this mitre are possible. Method A (see illustration) is the easier; in this case, a 56mm moulding is shown. Draw the line *ab* where the soffit and reveal mouldings intersect and this is the mitre. Method B is geometrical, and although not always adopted, is a useful one. The setting out is as follows:

Draw the springing line, mark in the rise and set out the intrados curve in the manner previously described. Draw a line to connect the springing and striking points AB. From point B draw a line at 90°, and bisect the resulting angle between this line and the reveal line; this is the mitre required.

Notice the arrangement of skewback voussoir and the moulded reveal brick necessary for proper construction.

Figure 8.34 shows a label course constructed over a drop Gothic arch.

Figure 8.35 shows a 'Welsh arch,' a useful method of covering ventilator openings or for rainwater outlets.

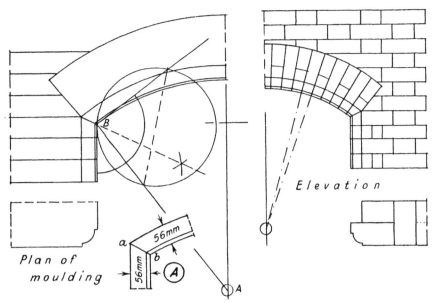

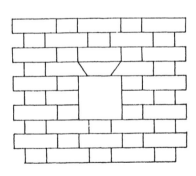

Elevation

Plan of moulding

Figure 8.33 Moulded segmental arch

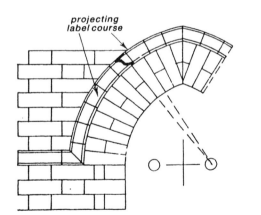

projecting
label course

Figure 8.34 Drop gothic arch with moulded label course

Figure 8.35 Elevation of wall showing a Welsh arch

Cutting an axed arch

Setting out

If an axed arch is to look right, each voussoir must have exactly the same tapering shape, so that mortar joints between them are parallel.

It is now common practice for brick manufacturers to calculate the necessary degree of taper, using CAD (Computer Aided Design) machines, when these arch voussoirs are to be supplied as moulded, specially shaped bricks.

Traditionally, axed brick arches have been produced by cutting bricks using hammer and bolster, after making a pattern templet of the voussoir shape required. This templet is made from a piece of timber 6mm to 12mm thick, having first drawn the arch (or half of its elevation) full size on a sheet of plywood.

The craft processes of setting out an axed arch to produce a templet, and then checking it by 'traversing', are dealt with in the following pages.

The tools required are:
1. Trammel heads; 2. dividers; 3. bevel; 4. traversing rules; 5. measuring rule; 6. straight edge; 7. carpenter's tools for cutting wood templet.

It is only necessary to set out half the arch, to include the key brick. For the purposes of this description, a segmental arch has been chosen.

Draw the centre line and set off the springing line at right angles to this. Ascertain the springing point and draw intrados, extrados, and skewback (Fig. 8.36). Set the dividers to the type of brick being used, and mark out voussoirs on the extrados. Having once set the dividers, do not open them any further, but, if necessary, close them a little. If opened, and a templet is made to the resulting voussoir shape, the bricks will not 'hold out' or will be too narrow to use. Mark out the templet to project approximately 50mm above extrados and 150mm below intrados. The templet can be made from a piece of wood 6mm to 12mm in thickness, and cut to a taper.

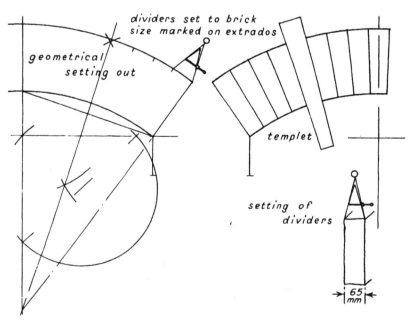

Figure 8.36   Setting out an axed arch

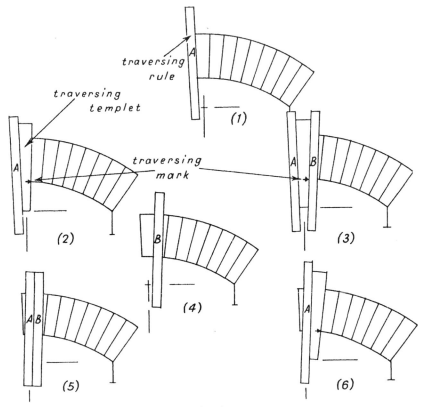

Figure 8.37   Traversing a templet for an axed arch

Traversing the arch

Making the templet shape, as described above, gives only the approximate size. To obtain the true shape, it must be traversed or traced over the face of the arch. Traversing is simply a way of multiplying any slight error in the tapered timber shape. To do this, follow Fig. 8.37.

1. Place the traversing rule A to the key brick.
2. Arrange the templet to fit a voussoir and mark a line on the side of the templet to coincide with intrados. This is called the 'traversing mark'.
3. Place the traversing rule B.
4. Remove A and the templet.
5. Place A to B
6. Remove B and again fit templet, allowing traversing mark to coincide with intrados.

Repeat these operations until the skewback is reached.

If the templet 'fills in' faster at the extrados than at the intrados, i.e. if it touches the top of the skewback but does not reach the bottom, it must be made smaller at the top by being planed down. The reverse applies if the templet fills in faster at the intrados. If, having been traversed, the templet is parallel to the skewback but fails to reach it, or 'under-fills' the area

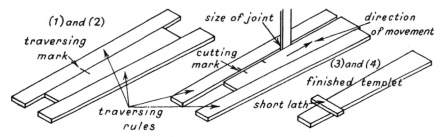

Figure 8.38   Applying the cutting mark to the templet to give joint allowance between voussoirs

between the key and the skewback, the templet must be brought down by placing the traversing mark higher up the templet. If the templet 'over-fills,' then the traversing mark must be lowered. The operation of travers-ing is repeated until the templet fits exactly between the key and the skewback, having in mind the number of voussoirs first required in the setting out drawing.

Traversing is an important operation in the construction of arches. Having obtained the correct shape of the templet, the voussoirs cut to this templet will find their true place in the arch.

Making the joint allowance

Allowance must be made for the mortar joint, and the cutting mark is obtained as follows (Fig. 8.38):

1. Place the templet between two traversing rules.

2. Mark a line on one traversing rule to coincide with the traversing mark on the templet.

3. Holding the traversing rules firmly, tap the templet up until the gap between the traversing rule and the templet gives approximately a 10mm joint.

4. Transfer the mark on the traversing rule (which was the traversing mark on the templet) to the templet; this is the cutting mark. Ascertain the bevel (Fig. 8.39), remembering that the bevel is at a tangent to the curve of the arch and is found by taking the mean of the squares drawn from both sides of the templet. Fix a short length of lath to the templet at this bevel and the templet is ready for use.

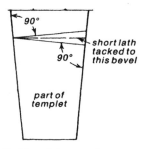

Figure 8.39   Squaring from both sides of the templet

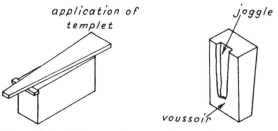

application of
templet

joggle

voussoir

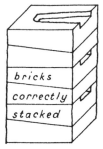

bricks
correctly
stacked

Figure 8.40 Using templet to mark voussoirs
Figure 8.41 Joggling voussoirs with comb hammer

Figure 8.42

Cutting arch voussoirs

Consideration must be given to the choice of facing bricks for axed arches. Bricks should only be selected that can be conveniently cut with a hammer and bolster into a tapering shape, without excessive wastage.

Mechanical masonry bench saws can be used for this cutting work with harder or perforated bricks. Class A and B engineering quality bricks will be very difficult to cut by any method, and are really unsuited to axed arch construction. If an arch of engineering quality Class A or B bricks is needed, they are best produced by the brick manufacturer as moulded or special shaped bricks.

Cutting. The tools required are:
1. hammer; 2. bolster; 3. pencil; 4. scutch; 5. comb hammer; 6. templet; 7. a piece of carborundum, to be used to give a clean, sharp arris to the voussoirs.

Apply the templet (Fig. 8.40), scribe with a pencil, and proceed with the cutting. When finished to the required shape, cut a joggle (Fig. 8.41), which allows the arch to be grouted when set in its position, giving added security. (Grout is a mixture of neat cement and water made into a thin slurry.) Stack the bricks carefully when cut (Fig. 8.42). A well-cut arch should stack evenly, each brick fitting snugly to the one above and below it.

Cutting an axed camber arch

The cutting of the camber arch is different from any other arch in that the bevels are not at a tangent to the same curve and are therefore all different. Some bricklayers consider this arch to be the most difficult to prepare and cut, but if the operations are carried out systematically, no difficulty should arise.

Preparing the templet

Set out the arch as previously described; only half the arch is necessary for practical purposes. There are two methods of drawing the camber:
1. Insert a small tack at the springing points and the top of rise respectively. Spring a lath (Fig. 8.43) arch between these points and draw a pencil line round it. This method gives a suitable curve, and although it

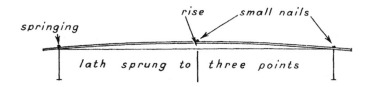

Figure 8.43   Single lath to draw camber soffit

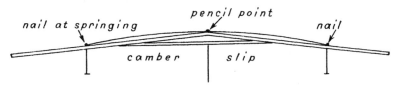

Figure 8.44   Three laths tacked together to draw camber soffit

cannot be used on an arch with a deep rise, it is quite adequate for a camber arch.

2. The camber slip method (Fig. 8.44) gives a true curve and is suitable for an arch with any rise. Construct a triangular framing of laths, insert a tack at each of the springing points, and place the framing to these tacks. The apex of the triangle must be at a distance equal to the rise, above the springing line. Place a pencil at the apex of the framing and trace it round, always keeping the arms of the triangle in contact with the tacks at the springing points.

Mark the voussoirs on the extrados, drawing in the key brick shape only, for the time being. To obtain the approximate size at the intrados, divide the distance between the key and the skewback into the same number of parts as there are voussoirs in the extrados. Make the approximate sized templet and traverse as previously described. When the correct size of the templet has been finally reached, traverse again, marking in the joint lines whilst doing so (Fig. 8.45). This will give the precise size of the voussoirs at the intrados, having in mind that each is slightly different. The obtaining of

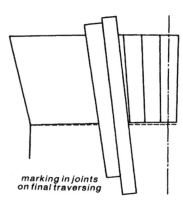

Figure 8.45   Drawing joint lines

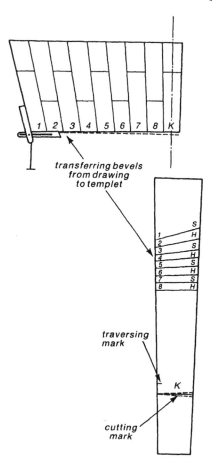

transferring bevels
from drawing
to templet

traversing
mark

K

cutting
mark

Figure 8.46

the correct size is important for purposes of setting or fixing the arch, as will be described later.

An alternative method of obtaining the joint lines is to produce the skewback lines downwards towards the centre line of the arch, thus obtaining a striking point from which to mark the joint lines. This would, however, be at a considerable distance from the springing line and the method shown above is much more convenient.

The final preparation of the templet differs from that required for the ordinary arch. As already stated, each brick has a different bevel, which must be cut on the brick before it is reduced to its voussoir size. It will be noted that the natural face of the brick can be maintained on the soffit of a normal arch, but with the camber arch this is impossible. To complete the preparation of the templet, therefore, transfer all the soffit bevels from the drawing to the templet, numbering from the skewback to the key (see Fig. 8.46). If the arch is bonded, as in a 300mm face, letter 'H' or 'S,' header or stretcher, according to the brick's position on the soffit. All the information needed will then appear on the templet and this is a wise precaution as a drawing may be lost or obliterated. Note that the practice of marking the bevels on the templet is for convenience only. Remember that the templet is wedge-shaped, so that if the bevels are placed on the templet from the right-hand side, they must be taken off in a similar manner.

Cutting

The camber arch has right- and left-hand sides, again differing from the ordinary arch, where a voussoir will take any position on its own curvature.

Consider the cutting of an arch with a 300mm face. In addition to the voussoir templet, two others are required (Fig. 8.47). These are called the 200mm and 100mm templets. The arch must be cut to a definite system. For instance:

1. Cut the soffit bevels on all the bricks, stacking as shown in Fig. 8.48.
2. On the back face of each brick, scribe its particular number so that it will be placed in its correct position.
3. Reduce the bricks to their voussoir size.

In operation 1 above, apply the bevel which has previously been set to its number, scribe with a pencil to give a clear cutting line, and cut. To obtain the header face, apply the 100mm templet, and for the stretcher face apply the 200mm templet (Fig. 8.49).

In operation 3, when ready to reduce the bricks to voussoir size, fix the templet to a rigid flat base. It will be found convenient to make the fixings coincide with the cutting marks (Fig. 8.50). Place the bricks to the templet, scribe, and cut. Stack neatly in the manner illustrated (Fig. 8.51).

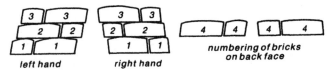

100mm and 200mm templets

Figure 8.47   Header and stretcher templets for camber arch

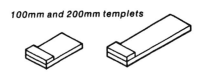

left hand   *right hand*   *numbering of bricks on back face*

Figure 8.48   Camber arch voussoirs stacked by numbers

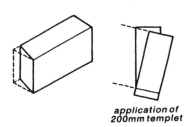

application of 200mm templet

Figure 8.49   Marking stretcher voussoirs

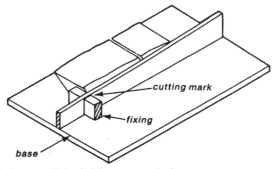

Figure 8.50 Reducing or rubbing bricks to voussoir size

Figure 8.51 Stacking finished voussoirs

Setting or fixing of arches _____

Types of support

No attempt will be made under this heading to show the type of temporary arch supports over large spans. Only those met with by the bricklayer in the course of everyday work will be dealt with.

Turning piece is made from solid timber for arches of limited span. For an arch with a 225mm soffit, two 75mm timbers can be placed side by side (Fig. 8.52). Alternatively, proprietary turning pieces can be supplied, made from solid expanded polystyrene.

Open-lagged centre is framed from light timber and used for the turning of ringed arches (Fig. 8.53).

Close-lagged centre is suitable for the turning of an axed or gauged arch. Its use facilitates the marking of the voussoir positions (Fig. 8.54).

When using temporary arch supports, folding wedges are placed directly under the centre. This facilitates the removal of the centre or turning piece and, in particular, avoids the chipping of the arch face on the intrados arris, which invariably occurs if wedges are not used.

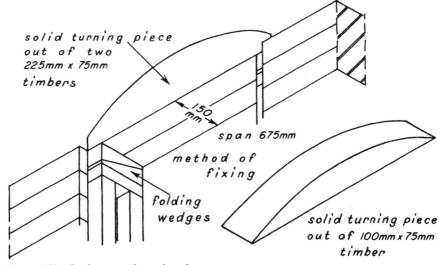

solid turning piece out of two 225mm x 75mm timbers

150 mm

span 675mm

method of fixing

folding wedges

solid turning piece out of 100mm x 75mm timber

Figure 8.52   Setting up arch turning pieces

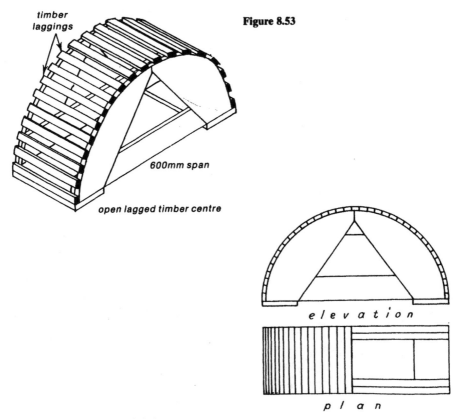

timber laggings

Figure 8.53

600mm span

open lagged timber centre

elevation

plan

Figure 8.54   Close lagged timber centre

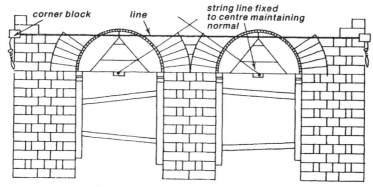

methods adopted to maintain alignment of arch

Figure 8.55 Setting arches

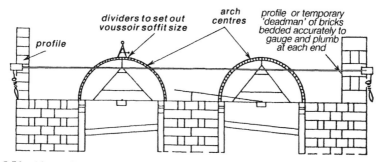

Figure 8.56 Alternative method to maintain alignment of arches

Preparation for setting

Consider the setting of a semi-circular arch:

Fix and adjust the temporary arch support. At the striking point on the centre fix a short length of bricklayer's line for checking that every voussoir is radiating correctly to the striking point when it is bedded (Fig. 8.55). With a pair of dividers, mark on the arch centre the correct positions of all the voussoirs.

To keep the arch in alignment, various methods can be adopted, e.g. if the arches are on a small flank wall, as in Fig. 8.55, the corners can be erected, using line and pins. If, on the other hand, the arch is one of many occurring on a large frontage, then a good method is to use profiles as Fig. 8.56.

For the cutting of skewbacks in preparation for the fixing of axed segmental or camber arches, always use a 'gun' or templet. Its application is shown in Fig. 8.57, and its function is to keep the angle of the skewback constant. It often happens that several bricklayers are engaged in building a long frontage and each is required to cut the skewbacks in a particular stretch of walling. If each is given the set-back of the skewback and then left to work independently, errors will occur, but if provided with 'guns' previously prepared by the foreman bricklayer and used in the proportion

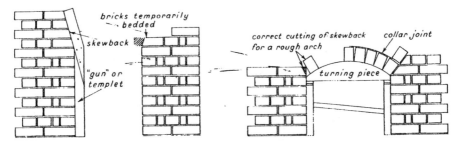

Figure 8.57 Use of gun templet to obtain correct skewback angles

Figure 8.58 Alternative method of marking skewback angle for rough arch

of, say, one gun to each group of three or four bricklayers, errors will be avoided.

For the cutting of a skewback for a ringed arch no gun is needed, as the turning piece can be fixed and the skewback ascertained from this by placing a brick flat on the turning piece, as shown in Fig. 8.58.

Setting a camber arch

Having cut the skewbacks and fixed the turning piece, preparation is made for the marking of the voussoir positions. To do this, take a lath and place it on the drawing or setting out of the arch, transfer the soffit marks to the lath, and apply to the turning piece, working right and left from the centre (Fig. 8.59). Note the reason for additional traversing to obtain precise voussoir sizes, as previously described. Use the line and pins (Fig. 8.60) to maintain alignment. In cutting a camber arch, the top portions of bricks occurring on the extrados are often uncut and adjusted after being fixed or set, see point A in Fig. 8.60. Note that the soffit of the arch is cambered but that the extrados is horizontal so as to conform with the general brickwork.

Special attention should be given to the bond setting out of general brickwork where a different coloured dressing occurs on the reveals and a camber arch is used to cover an opening. Most architects prefer the dressings to finish at the top of the arch, as shown in Fig. 8.61. The appearance is rectangular and not as illustrated in Fig. 8.62. The arrangement of bricks at the setting out level, which will probably be at ground level, must be watched closely so that proper sequence is obtained.

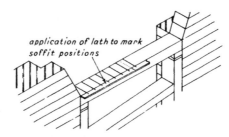

Figure 8.59 Marking voussoirs on camber arch support timber

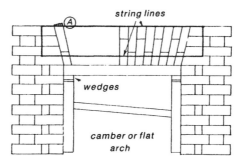

Figure 8.60 Setting camber arch voussoirs with string lines top and bottom

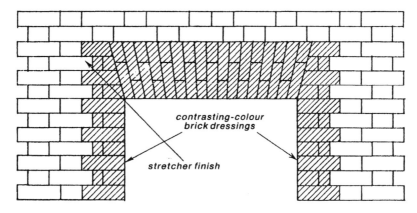

Figure 8.61 Overall rectangular finish of dressing must be anticipated when setting out facework at ground level

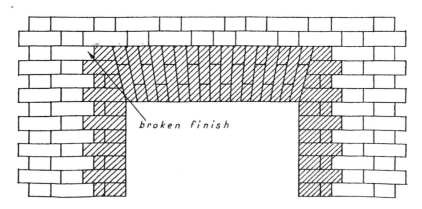

Figure 8.62 Alternative appearance to Fig. 8.61

Bull's-eye

(Figures 8.63 to 8.65). Although the detailed study of the construction of a bull's-eye belongs to the more advanced stage of brickwork, its construction will be dealt with in this chapter.

The setting out and cutting of an axed bull's-eye and the setting or fixing of this and the ringed bull's-eye differ slightly from the ordinary arch in that the lower half or invert requires no temporary supports, but needs the aid of a trammel for its formation in the wall (lower half of Fig. 8.63). Note that the key brick of the invert is laid first and that the voussoir positions must be marked off from its extrados, and not marked off from its intrados, on the centre or turning piece, as in the case of the arch.

The illustrations which follow will introduce some craft terms. From the information already given, the apprentice will be able to follow the craft operations involved in their erection.

Intersecting semi-circular arch ⎯⎯⎯⎯⎯⎯

(Figure 8.66). Gauged half-brick intersecting semi-circular arch, with moulded intrados to conform with moulded reveals. There is a half-brick moulded sill made up to courses by the introduction of a tile course.

Figure 8.67 shows the intersection of two semi-circular arches, 225mm on face, on a 330mm pier. Note the cutting of the brick in the second course from springing.

Arch with a brick core or tympanum ⎯⎯⎯

The brick core can be erected in two ways:

  1. By the use of a trammel, using the core as a support to the main arch during construction.

  2. By first erecting the main arch on a temporary support or centre and inserting the smaller arches and brick filling when the temporary support is removed (Fig. 8.68). Two examples are shown in Figs. 8.68 and 8.69.

Indented arch ⎯⎯⎯⎯⎯⎯⎯⎯⎯⎯

(Figure 8.70). Sometimes termed a 'blocked' or 'rusticated' arch. The block on reveals and arch are usually in a different coloured brick, an attractive combination being that of a sand-faced multi-red brick for the blocks with a silver-grey facing brick for the general brickwork. The courses between each block are in the same colour as the general brickwork.

Note the method of obtaining correct sizes of voussoirs, by constructing a curve concentric to the intrados from the greatest depth of arch face.

Cant brick arch ⎯⎯⎯⎯⎯⎯⎯⎯⎯

(Figure 8.71). Shows the method of cutting axed arches constructed of cant or bullnose bricks.

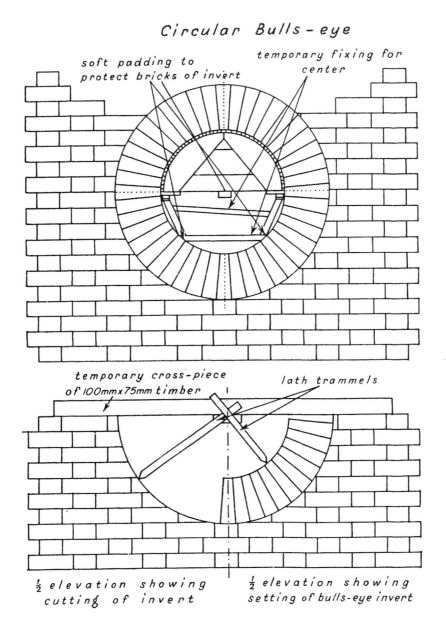

Figure 8.63 Circular bulls-eye

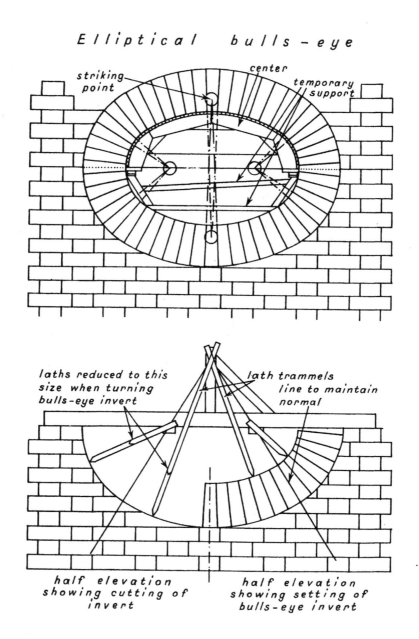

Figure 8.64   Elliptical bulls-eye with major axis horizontal

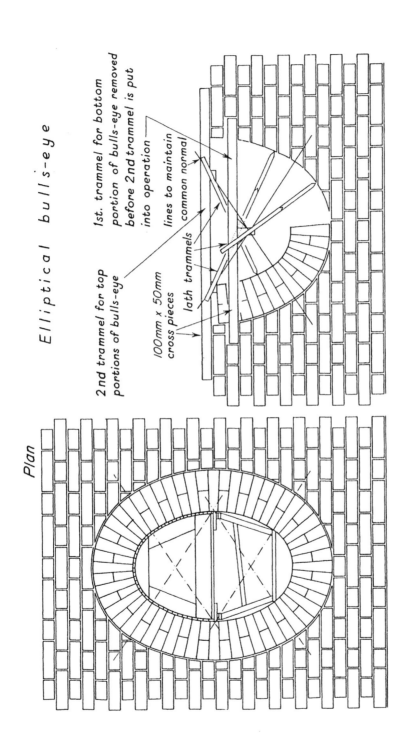

Figure 8.65 Elliptical bulls-eye with major axis vertical

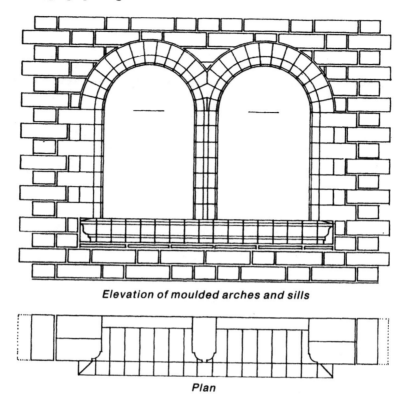

Elevation of moulded arches and sills

Plan

Figure 8.66   Intersecting moulded arches

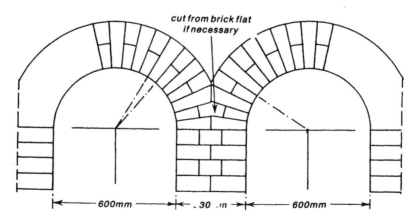

cut from brick flat
if necessary

|← 600mm →|← 30 .m →|← 600mm →|

Figure 8.67   Intersecting axed arches, 225mm bonded face

Arches with basketweave brick core or tympanum

A

elevation

section

Figure 8.68

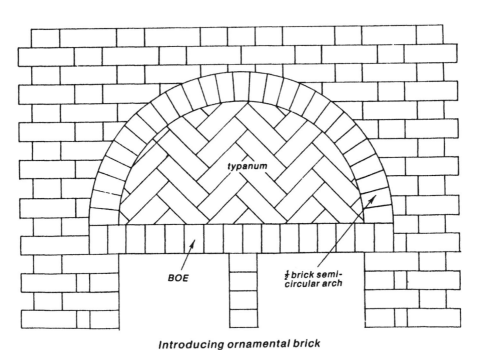

typanum

BOE

½ brick semi-circular arch

*Introducing ornamental brick
filling or tympanum*

Figure 8.69 Introducing ornamental brick filling or tympanum

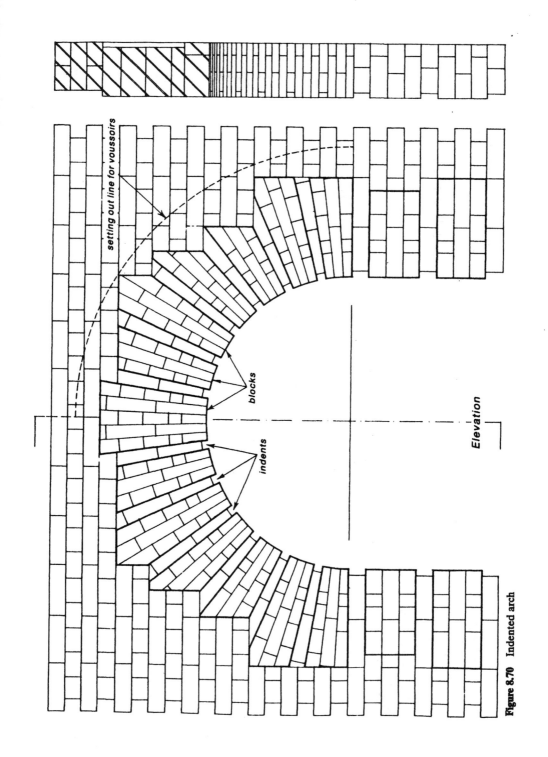

Figure 8.70 Indented arch

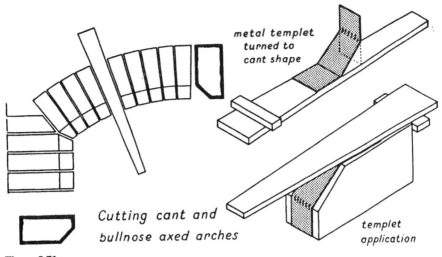

metal templet
turned to
cant shape

Cutting cant and
bullnose axed arches

templet
application

Figure 8.71

Stilted arches

(Figure 8.72). Although this type of arch was used in isolated cases for decorative purposes, it can be and was originally used to maintain alignment at the crowns of arches where small openings occurred together with large openings.

Horse-shoe arch

(Figure 8.73). The origin of this arch was probably oriental and the crown of the arch was often pointed instead of circular, and in such a case the arch was known as 'Moorish.'

Arches in orders

In these the brickwork was recessed in 'orders,' as shown in Fig. 8.74, and in the early days they occurred in very thick walls. Today, orders do not necessarily occur in thick walls, and can be created for decorative purposes in external cavity wall construction. A probable reason for their use in early building was the difficulty of obtaining and constructing temporary supports: by recessing, only a temporary support for the inner arch ring was necessary, successive orders being corbelled. In modern practice, where the orders project 112.5mm, it is advisable to use temporary supports for every projection.

This type of arch was originally adopted to overcome difficulties in construction. Other forms of opening also occurred in thick external walls; the width on the external face might be narrow while in the internal face it was broad, often with splayed jambs. Several reasons are given for their

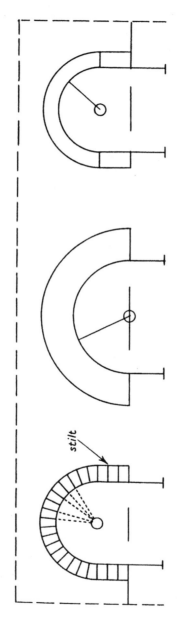

Figure 8.72   Use of stilted arches to make crowns level with larger span arch

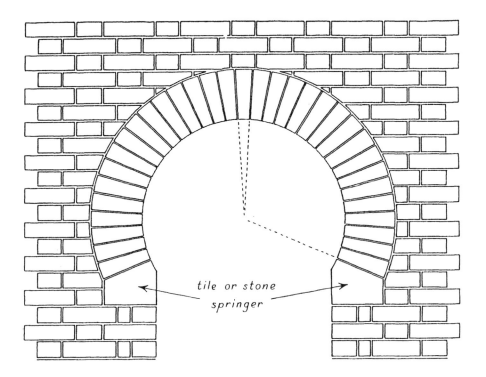

tile or stone
springer

Figure 8.73 Horse-shoe arch

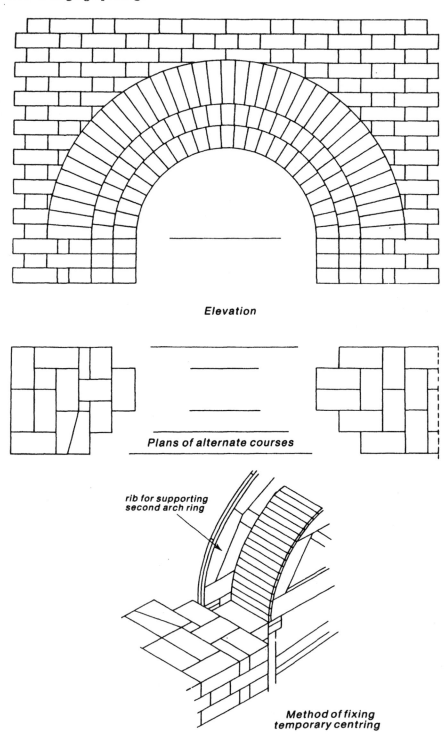

Elevation

Plans of alternate courses

rib for supporting
second arch ring

*Method of fixing
temporary centring*

Figure 8.74   Construction of recessed arches

use; for instance, a narrow arch on the external face of the wall would prevent the entry of thieves, glass being either unknown or not available, while a wide arch on the inner face would give light; again such orders were constructed in castles to protect the archers withstanding a siege.

Arches in orders on splayed jambs

Fig. 8.75 illustrates this type of arch. Care must be taken when erecting the temporary ribs used for the turning of the separate rings to see that these can be easily removed after construction. If notches are not cut at the springing of the ribs where they rest on the recess, it is obvious that they will bind and will be difficult to remove. This will tend to spall the edges of the arch soffit and also the top of the brick jamb. The recess occurring between the arch and jamb should be weathered by some suitable method, e.g. by tile creasing built in or by continuance of the jambs until they intersect the soffit of one ring of the arch and the face of the other.

Rear-arch

Figures 8.76 to 8.79 show the plan, elevation, and section of a rear-arch. This is a segmental pointed arch and it will be noted that the jambs of the opening are splayed but the soffit of the arch is level. It is considered that the best method of construction is shown in Fig. 8.76 (Method 1). The skewback is cut through the abutment and the splayed jamb is 'made good' to the soffit of the arch after it has been built and the temporary centering removed.

Fig. 8.77 (Method 2) might be adopted, and in this case the intersection between the splayed jambs and the level soffit of the arch must be developed in order to gain the true shape of the templet for cutting the skewback. Fig. 8.78 illustrates the method of obtaining the true shape of intersection. The length of the splayed jamb in plan is divided in this case into eight equal parts and the points projected vertically into the elevation. The distances from the line *xy*, i.e. *a*0 ... *b*1 ... *c*2 ... *d*3 ... etc., are taken from the elevation (Fig. 8.79).

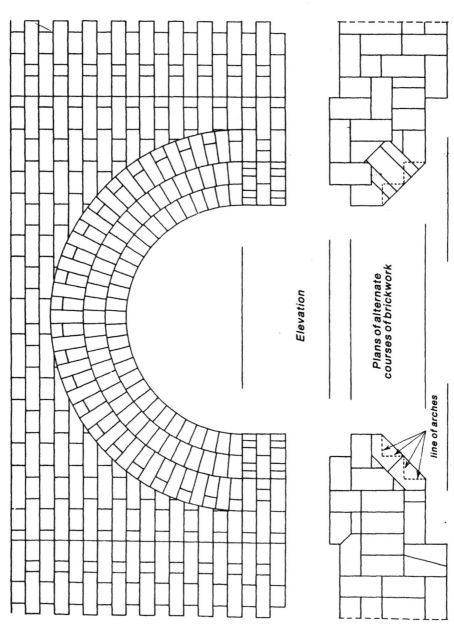

Elevation

Plans of alternate courses of brickwork

line of arches

Figure 8.75   Recessed arches on splayed jambs

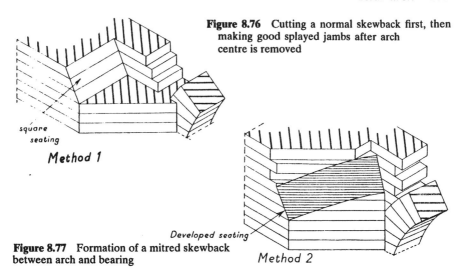

Figure 8.76 Cutting a normal skewback first, then making good splayed jambs after arch centre is removed

square seating

Method 1

Developed seating

Figure 8.77 Formation of a mitred skewback between arch and bearing

Method 2

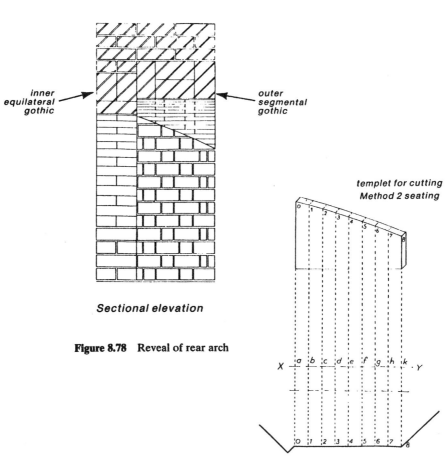

inner equilateral gothic

outer segmental gothic

templet for cutting Method 2 seating

Sectional elevation

Figure 8.78 Reveal of rear arch

X — a b c d e f g h k — Y

True shape of intersection

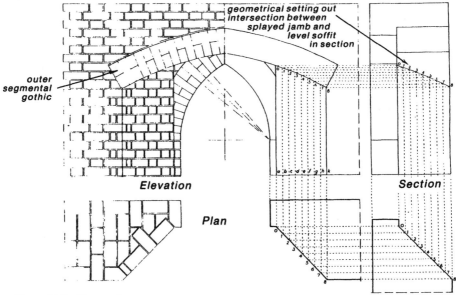

geometrical setting out
intersection between
splayed jamb and
level soffit
in section

outer
segmental
gothic

Elevation

Section

Plan

Figure 8.79   Rear arch

Bonnet arch _____

Figure 8.80 shows an axed arch where tiles have been introduced on the inner arch because of the thin nature of voussoirs that would have occurred at this point. A sound method to adopt in cutting this arch is to mount the templet for reducing bricks to voussoir shape on a flat board. Procedure for setting out and cutting:

1. Set out the face templet in the usual way (Fig. 8.81).

2. Prepare the templet for mounting, noting that the templet for the splay is elongated and that the points A and B in Fig. 8.82 must equal points *c* and *d* in Fig. 8.81.

3. Mount the templet and prepare bricks for reducing Fig. 8.83. The squint brick must be bevelled to the arch curvature before being reduced; other bricks can remain square. Having placed the bricks to the mounted templet, scribe and cut.

It will be realised that the arch, besides being wedge-shaped on the face and splayed soffit, is also wedge-shaped in its thickness, and the method of cutting as explained neglects allowance for this. In spite of this, the method described is simple and avoids complications, the size of the wedge in thickness of the arch is very small and the difference can be made up in mortar on the back edge.

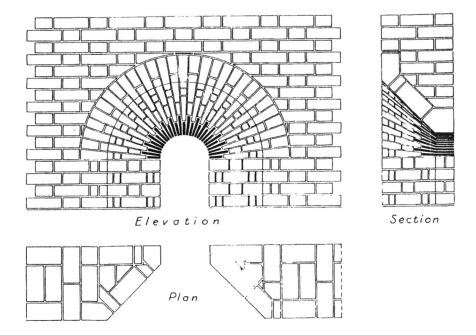

Elevation   Section

Plan

Figure 8.80   Bonnet arch

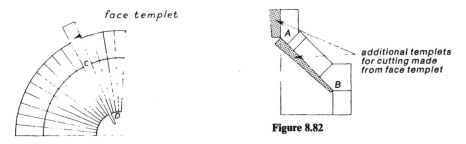

face templet

additional templets
for cutting made
from face templet

Figure 8.82

Figure 8.81   Elevation of bonnet arch
showing tapering voussoirs on face

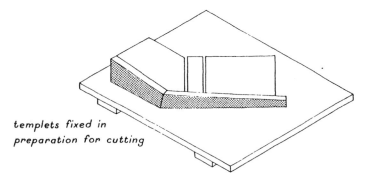

templets fixed in
preparation for cutting

Figure 8.83   Fig. 8.82 templets fixed in preparation for cutting voussoirs

9

Cavity walling and brick cladding

Definition

The standard form of construction for the external walls of brick buildings is called cavity walling. This means that the bricklayer builds the two separate 'leaves' or 'skins' of 'brick masonry', (a general term indicating brickwork and/or blockwork), with a 50mm to 75mm wide space between. The outer skin is usually 102.5mm thick face brickwork, but may be constructed from facing quality blocks. The inner skin is usually 100mm thick common blocks that are later plastered to receive internal decoration. (see Fig. 9.1).

Both skins of brick masonry are joined together with a regular pattern of corrosion resistant ties, so that they behave as one single wall.

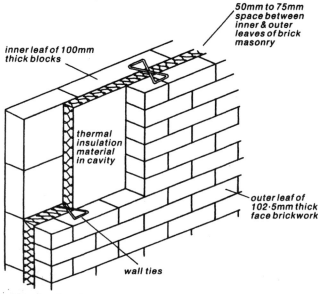

Figure 9.1 Typical insulated cavity walling

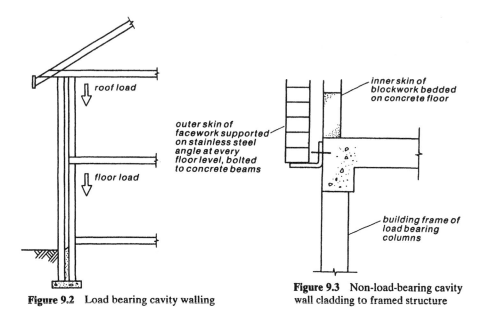

Figure 9.2 Load bearing cavity walling

Figure 9.3 Non-load-bearing cavity wall cladding to framed structure

Reason for the title _____

The reasons for the double title of this chapter are that from the bricklayer's point of view, the craft skill requirements for cavity wall brickwork and brick cladding are very similar.

Cavity wall brick masonry may be load bearing, that is, used to support the load from floors and roof. This application is usually confined to low rise buildings of up to three to four storeys only. (See Fig. 9.2).

Alternatively identical looking cavity walling is widely used to cover the framed structure of high rise/multi-storey building, where it is *not* required to support the load of floors and roof, and is referred to as brick cladding. (See Fig. 9.3).

Thirdly, a 102.5mm thick wall, (single skin), of face brickwork may be used to cover large pre-cast concrete panels of some types of industrial buildings or *in-situ* concrete walling and bridge parapets.

Fourthly, a single skin of 102.5mm face brickwork is used as brick cladding to timber frame houses. Here the inner leaf is formed from factory-made storey height panels of structural timber which support floor and roof loads. All the craft skill requirements of the bricklayer contained in this chapter refer equally to these four different applications of cavity walling.

Purpose _____

Cavity wall construction began to be widely used from the 1920s, as a way of preventing dampness from soaking through the outer walls of buildings.

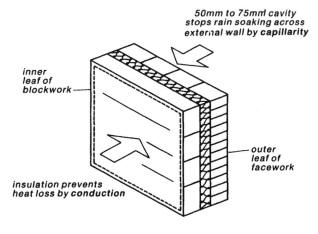

50mm to 75mm cavity
stops rain soaking across
external wall by capillarity

inner
leaf of
blockwork

outer
leaf of
facework

insulation prevents
heat loss by conduction

dotted area = 1m²

Figure 9.4   Function of typical cavity walling

The 50mm to 75mm wide gap stopped rain penetrating from the outer surface to the plastered inner surface of external walls by capillarity (when water is drawn through hairlike channels within a porous structure, a brick say, by the action of surface tension). This was possible when outer walls were commonly 215mm thick solid brickwork.

As a secondary advantage, this space between the inner and outer skins also provides thermal insulation for modern buildings, because heat energy mainly escapes by conduction through solid material. Air is a poor conductor of heat energy, therefore the rate of heat loss is very much slower than was the case when buildings had solid outer walls, (see Fig. 9.4).

Building regulations

The basic advantages of cavity wall construction, over solid 215mm thick brickwork, for the outer walls of a building have been incorporated in the Building Regulations. In order to satisfy requirements of the Building Regulations and enable planning permission to be obtained, cavity wall construction is usually specified whether the structure is low rise or multi-storey.

Figure 9.5 indicates how basic requirements of the Building Regulations are satisfied, where standard strip foundations are specified with a solid ground floor slab.

Figure 9.6 shows a trench-fill foundation, associated with a suspended ground floor construction of pre-stressed, precast concrete floor beams supporting standard size concrete blocks.

A   External cavity wall, providing resistance to through-penetration of rain

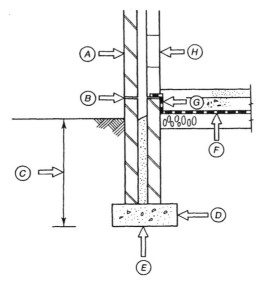

Figure 9.5 Cavity walling on standard strip foundation indicating Building Regulation requirements with solid ground floor

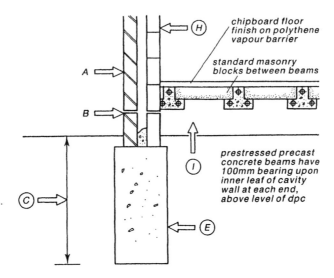

chipboard floor finish on polythene vapour barrier

standard masonry blocks between beams

prestressed precast concrete beams have 100mm bearing upon inner leaf of cavity wall at each end, above level of dpc

Figure 9.6 Cavity walling on trench fill foundation with suspended concrete ground floor

B Horizontal dpc in both leaves, not less than 150mm above ground level, to prevent dampness rising from the soil
C A minimum distance of one metre between ground level and the underside of the concrete foundation, as a protection against frost heave in winter and drying shrinkage of clay subsoils in summer
D A minimum 150mm thickness of foundation concrete to transfer the building load adequately on to the 'natural foundation' of the subsoil

E   Sulphate resisting cement in the foundation concrete and substructure brickwork, necessary where soluble sulphates are present in subsoil water

F   Continuous damp proof membrane (dpm) across the ground floor areas to prevent dampness rising from the soil

G   Dpm and dpc, lapped and joined within the floor thickness around the perimeter of rooms

H   External walls built of lightweight blocks and other thermal-insulating material to give a 'U' value of 0.45. (In other words, heat energy must not escape through the outer walls at a rate greater than 0.45 watts per square metre, per hour, per degree difference in temperature internally and externally)

I   A ventilated air space separates the suspended floor from damp soil

Function

Fig. 9.2 shows that it is the inner leaf of cavity walling that largely supports the load from floors and roof in a low rise building of load-bearing wall construction. Common building *blocks* have totally replaced common *bricks* for this inner leaf, due to improved thermal insulation values and also bricklayer output (i.e. one standard size 100mm thick block is equal to six bricks).

The purpose of the outer skin of facework is to give the building a weather resistant and pleasant appearance, by selection from the wide range of colours and surface textures of bricks and facing blocks available to the designer.

Although cavity walling is constructed with properly finished, solidly filled mortar joints, it is expected that this outer skin will let rainwater soak through as far as the cavity. This is because mortar, bricks and blocks are porous to varying degrees. This through-penetration will be highest on those elevations of a building exposed to prevailing wet winds.

Where cavity walling is used as cladding to a high rise building the degree of exposure to wind driven rain increases with height.

Thermal insulation

When cavity walling was first used, it was common practice to install air bricks at the top and bottom, at intervals around the whole perimeter of the building, to ventilate this 50mm to 75mm wide space to remove damp air. Since the 1950s however, cavity walls have become sealed, with insulating material built in as work proceeds, to improve the thermal insulation value of the cavity space.

Two ways that a bricklayer may be told to install thermal insulation in cavity walling are shown in Figs. 9.7 and 9.8. The fully filled system is where flexible fibre 'batts' of insulation completely occupy the cavity space. The partial fill system is where stiffer 'boards' of insulating material half-fill the cavity, but retain a 25mm to 35mm air space as well. Both batts and boards are supplied in purpose-made sizes to fit neatly between layers of wall ties and are approx. 900mm in length.

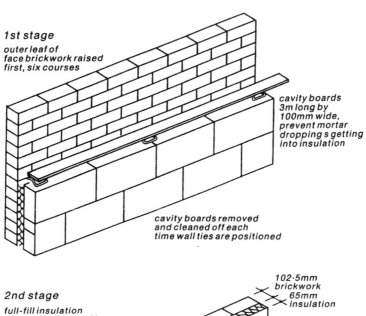

1st stage

*outer leaf of
face brickwork raised
first, six courses*

*cavity boards
3m long by
100mm wide,
prevent mortar
dropping s getting
into insulation*

*cavity boards removed
and cleaned off each
time wall ties are positioned*

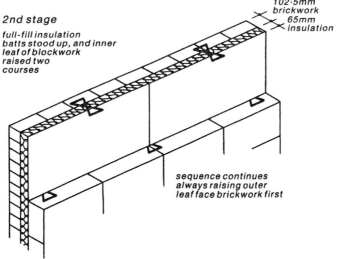

2nd stage

*full-fill insulation
batts stood up, and inner
leaf of blockwork
raised two
courses*

*102·5mm
brickwork
65mm
insulation*

*sequence continues
always raising outer
leaf face brickwork first*

Figure 9.7 Fully filled cavity insulation system

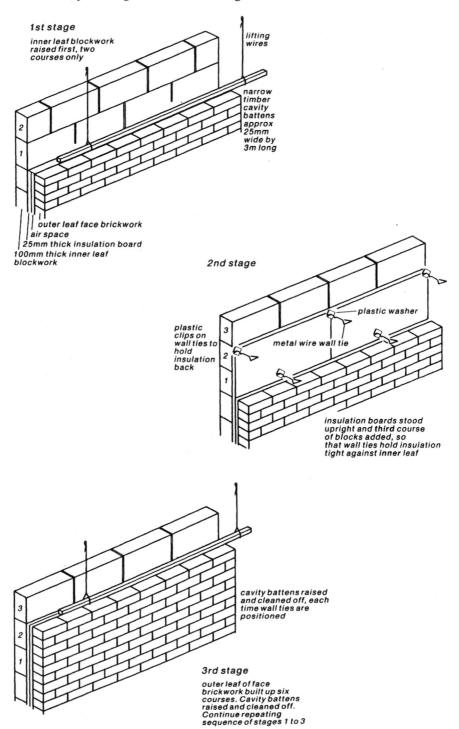

1st stage

inner leaf blockwork
raised first, two
courses only

lifting
wires

narrow
timber
cavity
battens
approx
25mm
wide by
3m long

outer leaf face brickwork
air space
25mm thick insulation board
100mm thick inner leaf
blockwork

2nd stage

plastic
clips on
wall ties to
hold
insulation
back

plastic washer

metal wire wall tie

insulation boards stood
upright and **third** course
of blocks added, so
that wall ties hold insulation
tight against inner leaf

cavity battens raised
and cleaned off, each
time wall ties are
positioned

3rd stage

outer leaf of face
brickwork built up six
courses. Cavity battens
raised and cleaned off.
Continue repeating
sequence of stages 1 to 3

Figure 9.8 Partially filled cavity insulation system

Bonding

Cavity wall construction has been responsible for the disappearance from modern buildings of the many and various traditional bonding patterns which are possible with the use of headers and stretchers (see Chapters 4 and 5). This is because use of the full range of face bonds require walls to be at least 215mm thick if headers are to be used effectively.

Stretcher bond is best suited to the 102.5mm thick outer skin of cavity walling and for making the most economic use of the longer stretcher face of each expensive facing brick, see Fig. 9.9. However if another face bond is required for cavity walling, to match existing work, then it is possible to use 'snapped headers', (half bats), as shown in Fig. 9.10.

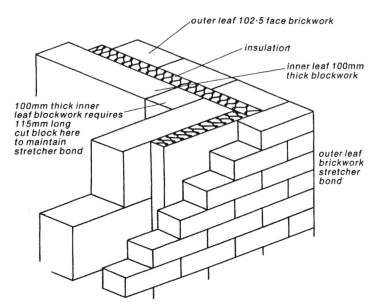

Figure 9.9 Cavity wall quoin, stretcher bond

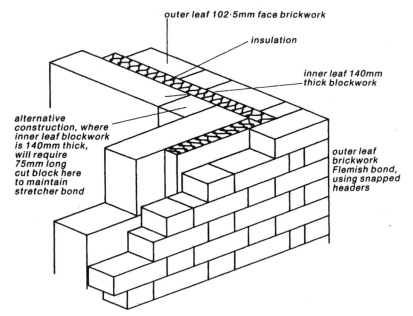

Figure 9.10   Cavity wall quoin, showing use of snap headers in 102mm outer leaf

DPC cavity trays

Bearing in mind that rain water will penetrate the outer skin of brickwork, trays or gutters of flexible bituminous felt must be built into cavity walls to collect the water and prevent dampness from reaching the inner skin of blockwork, see Fig. 9.11(a), showing patent steel lintel bridging an opening.

These cavity gutters of dpc material are needed above every opening in cavity wall construction and at every floor level of a multi-storey building, where cavity walling is used as the external cladding.

Purpose-made stop ends of dpc material—(see Fig. 9.11(b)—are glued at either end, to prevent run-off into wall insulation and to encourage draining through weep holes.

For alternative methods of supporting cavity work over openings, see Fig. 9.12.

'Running laps' at the end of one roll of dpc and the next, including internal and external angles, must be effectively sealed and jointed, so as to remain watertight.

Purpose made polyethylene units may be specified to provide permanent support to running laps in dpc cavity gutters, as shown in Fig. 9.15.

Rolls of dpc should always be stored *on end* to avoid squashing and distortion, which will make the material more difficult to flatten out when used. In cold weather, the rolls should be stored in a warmer place to make bedding the material that much easier.

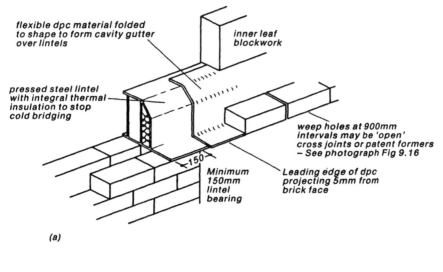

flexible dpc material folded
to shape to form cavity gutter
over lintels

inner leaf
blockwork

pressed steel lintel
with integral thermal
insulation to stop
cold bridging

weep holes at 900mm
intervals may be 'open'
cross joints or patent formers
– See photograph Fig 9.16

150

Minimum
150mm
lintel
bearing

Leading edge of dpc
projecting 5mm from
brick face

(a)

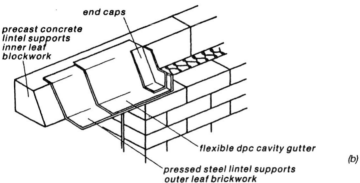

end caps

precast concrete
lintel supports
inner leaf
blockwork

flexible dpc cavity gutter

(b)

pressed steel lintel supports
outer leaf brickwork

Figure 9.11 (a) Typical cavity tray or gutter; (b) purpose-made end caps glued to cavity
tray

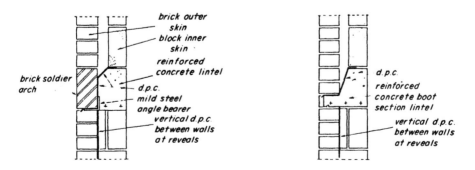

brick outer
skin

block inner
skin

reinforced
concrete lintel

brick soldier
arch

d.p.c.
mild steel
angle bearer
vertical d.p.c.
between walls
at reveals

d.p.c.

reinforced
concrete boot
section lintel

vertical d.p.c.
between walls
at reveals

Figure 9.12 Alternative methods of supporting brickwork across openings

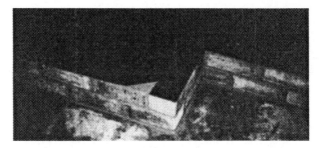

Figure 9.13   Lap jointing flexible dpc a minimum of 100mm in running laps between the end of one roll and the next, and also at corners, as shown. *Note* mortar bedding *under* dpc, which should be left flush with wall face or projecting 5mm

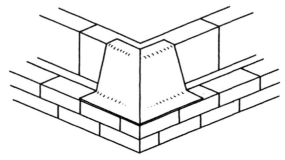

Figure 9.14   Purpose made dpc tray corner unit

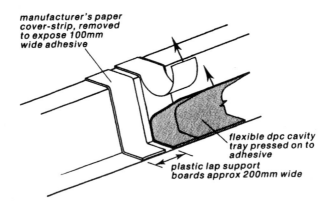

manufacturer's paper cover-strip, removed to expose 100mm wide adhesive

flexible dpc cavity tray pressed on to adhesive

plastic lap support boards approx 200mm wide

Figure 9.15   Patent rigid plastic support board for running laps in dpc trays

Weep holes _____

Empty or 'open' cross joints are left at intervals of 900mm in the outer leaf of cavity walling, at the level of cavity gutters, to act as permanent drainage points. These 'weep holes' may simply be left as a cross joint free of mortar, or formed with plastic or nylon fibre inserts. The latter have the advantage of acting as filters which will hold back any salts in water draining from the cavity that could create localised staining. See Fig. 9.16.

Both types of weephole inserts are available in a limited number of colours to match mortar joint colour. Where a light colour mortar is specified, open weep holes appear black and will be conspicuous if not in matching locations above and below every window opening. With the external elevations of a building, it is important that the bricklayer takes care to keep the pattern of weep holes under and over openings symmetrical.

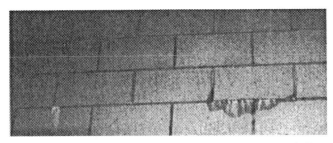

Figure 9.16 Soluble salt staining from cavity tray weep holes in external wall of dense facing blockwork. Excessive staining along bed joint is due to absence of temporary covering of open cavity in very wet weather during construction

Wall ties _____

Headers and courses of headers in *solid* brick walling are there to tie the wall from back to front. (See Chapter 4, Bonding Rule 2). In cavity wall constructions headers cannot be used for this purpose, because they would allow dampness to cross the cavity by capillary action.

Therefore, a range of proprietary ties are made for this job of tying together inner and outer skins of brick masonry, as 'substitute headers' so that both leaves behave as *one* wall. These wall ties are made from stainless or galvanised steel, or polypropylene so they do not provide a passage for moisture.

The general and important requirements for any pattern of wall tie are as follows (see Fig. 9.18).

A   It must be made durable and non-corroding;
B   it must have a central drip or twist to prevent water from tracking across the cavity;
C   when laying it, allow for not less than 50mm embedment in each leaf of the cavity walling;

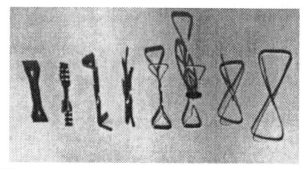

Figure 9.17   Different types of wall ties

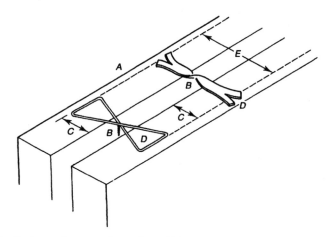

Figure 9.18   Basic requirements for cavity wall ties

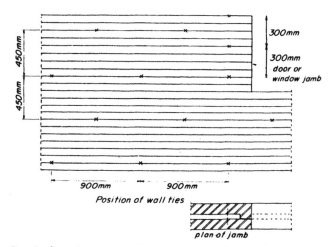

Figure 9.19   Standard spacing of wall ties (maximum distances apart)

D   it must have a particular end-shape, to give a secure grip in the bedjoint mortar; and

E   a tie of the correct overall length must be used, to span the cavity width plus two embedments.

The standard maximum spacing for wall ties is at intervals of 900mm horizontally and every sixth course vertically. Each horizontal layer should be offset (see Fig. 9.19). For purposes of estimating quantities of wall ties required, this works out at approx. 2.5 per m^2

Cleanliness

For cavity walling to be effective, wall ties, insulation and cavity gutters must be kept free of mortar droppings as work proceeds. If, due to carelessness and poor supervision, cavities are not kept clean, then dampness will be able to cross the cavity through porous mortar droppings.

Cavity battens or boards, approximately 3m long, raised and cleaned off every six courses as work proceeds, are the best way of preventing mortar droppings falling into cavity walling (see Fig. 9.20).

Alternatively, where fully filled cavity insulation is specified, plain battens without lifting wires are used, see Fig. 9.21.

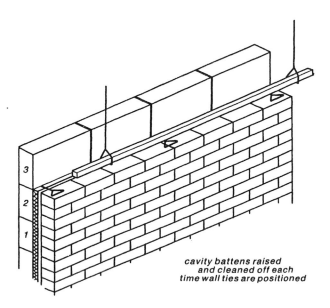

*cavity battens raised
and cleaned off each
time wall ties are positioned*

Figure 9.20   Cavity battens in use with partial-fill insulation

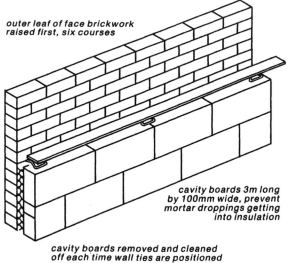

outer leaf of face brickwork
raised first, six courses

cavity boards 3m long
by 100mm wide, prevent
mortar droppings getting
into insulation

cavity boards removed and cleaned
off each time wall ties are positioned

Figure 9.21   Cavity battens in use with full-filled insulation

Coring holes

It is usual to leave temporary openings, called coring holes, in the outer
leaf, over all cavity trays. These holes, of one-brick size, are for the removal
of any mortar droppings, jointers, etc, that have got past the cavity battens,
see Fig. 9.22(A). Cavity ties and trays should be inspected and cleaned in
this way at the end of every day's work. Where possible, however, it is
better if the one-block sized coring holes be left out of the inner leaf, see
Fig. 9.22(B). This is recommended, as it provides a bigger temporary
opening. It also avoids the risk of the slight difference in mortar colour
which would highlight the coring holes in the finished facework. The coring
holes are sealed up when the external scaffolding is removed.

Temporarily, bedding bricks or blocks in sand at points A or B provide
support for those above, and make for easy removal to form the coring
holes.

Coring holes should only be regarded as a back-up procedure, and not a
substitute for cavity battens or boards.

Spreading mortar bed

An important feature when running out the bedding mortar for each
course of bricks for cavity work—which will considerably lessen the risk of
mortar dropping into the cavity—is shown in Fig. 9.23. If this technique is
maintained then there will be very little mortar to be recovered as each
brick is pressed into position.

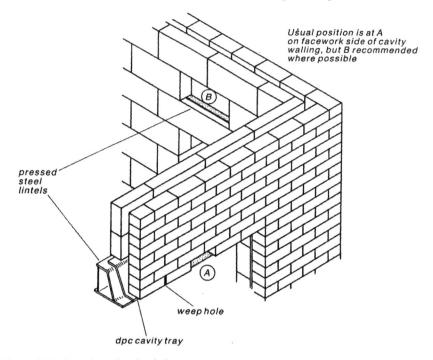

Usual position is at A
on facework side of cavity
walling, but B recommended
where possible

pressed
steel
lintels

weep hole

dpc cavity tray

Figure 9.22   Location of coring holes

Figure 9.23   Spreading the mortar bed in cavity construction, so as to minimise squeeze of mortar into the cavity side of the wall when each brick is pressed down to the line. (Note: furrow, full bed on face, and cavity-side mortar being splayed)

Openings

Leaving a door or window opening in a *solid* brick wall is a relatively simple job, as it merely requires making allowance for the correct space for a frame, plus a lintel or arch across at the right height.

Leaving the same opening in a *cavity* wall, however, is a much more complicated process. Detailed provision must be made for dealing with:

1 Dampness, which may soak through the sides or reveals of any opening (see Fig. 9.24)
2 Rainwater, which may run down inside the cavity from above the opening (see Figs. 9.11 and 9.15)
3 Rain, which may soak through any sill or threshold of an opening (see Figs. 9.25 to 9.29).

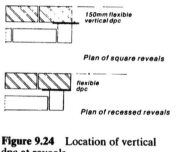

Figure 9.24   Location of vertical dpc at reveals

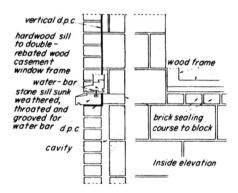

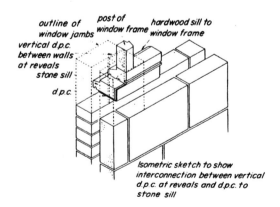

Figure 9.26   Construction at timber window frame on stone sill

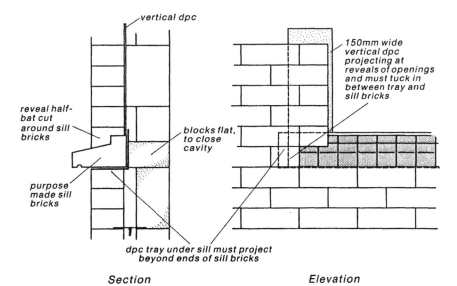

Figure 9.25 Purpose made sill bricks at window opening

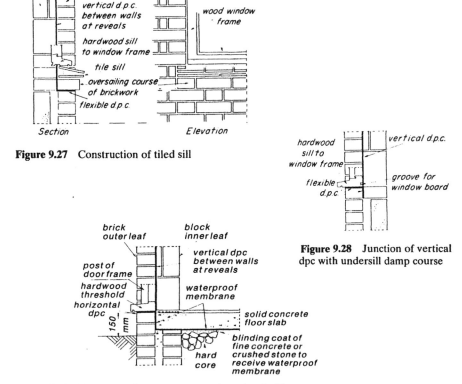

Figure 9.27 Construction of tiled sill

Figure 9.28 Junction of vertical dpc with undersill damp course

Figure 9.29 Damp proofing treatment at door threshold

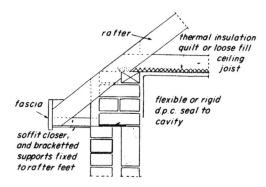

rafter

thermal insulation
quilt or loose fill

ceiling
joist

fascia

flexible or rigid
d.p.c. seal to
cavity

soffit closer,
and bracketted
supports fixed
to rafter feet

Figure 9.30   Sealing cavity at eaves level

Sealing at eaves level _____

To prevent birds that get into the roof spaces from nesting in wall insulation, it is usual to close or seal off cavity walls at the top. A course of headers or blocks laid flat is a convenient way of doing this. This also spreads the load of a pitched roof between both inner and outer skins of brick masonry, see Fig. 9.30.

Brick cladding _____

With multi-storey buildings, the total loading of all the floors would be too much for normal 275mm-wide cavity walling to support safely. In these situations, a permanent frame of reinforced concrete, or structural steel is erected first, before external cavity walling is built around it, see Fig. 9.3.

Support for brick cladding

Although the bricklayer builds this cavity walling in exactly the same way as for low-rise structures, when used on multi-storey buildings, both leaves of brick masonry must be supported at every floor level. This ensures that permanent pressure does not build up on the brick masonry at ground level. Cavity wall cladding to multi-storey buildings is described as non-load-bearing, because it does not carry the weight of floor slabs resting upon it.

Support for the outer skin of brick cladding using reinforced concrete edge beams, as shown in Fig. 9.31 and 9.32, has generally been replaced by stainless steel angles or brackets, as shown in Figs. 9.33 to 9.35.

You will notice that Figs. 9.31 to 9.35 inclusive show a soft-joint filler of polyethylene foam strip under the supporting steel angle or concrete toe beams. This very important compression joint must be provided under each supporting angle of stainless steel, or reinforced concrete toe beam, around the external perimeter of a building at every floor level.

The horizontal compression joint, completely clear of mortar, permits different rates of expansion and contraction (up or down) to take place without restriction, between the building and the brick masonry cladding, see Fig. 9.36.

Unless these horizontal compression joints are provided at every floor level, serious cracking will result when the outer brickwork tries to expand and/or the load-bearing structural frame shrinks slightly over the years.

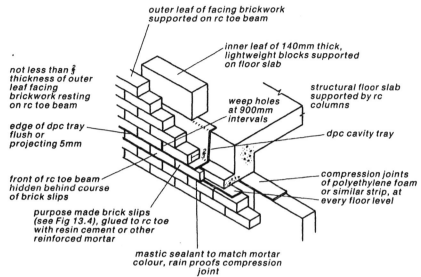

Figure 9.31 Support of cavity wall brick cladding on reinforced concrete toe beam

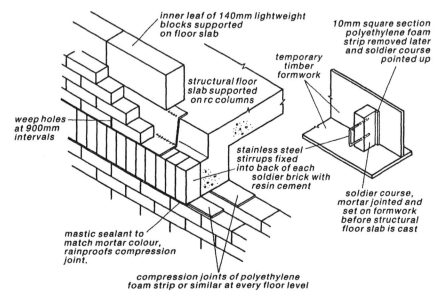

Figure 9.32 Outer leaf of cavity wall facework supported on soldier course course of bricks, cast *in situ* with RC structural floor slab and edge beam

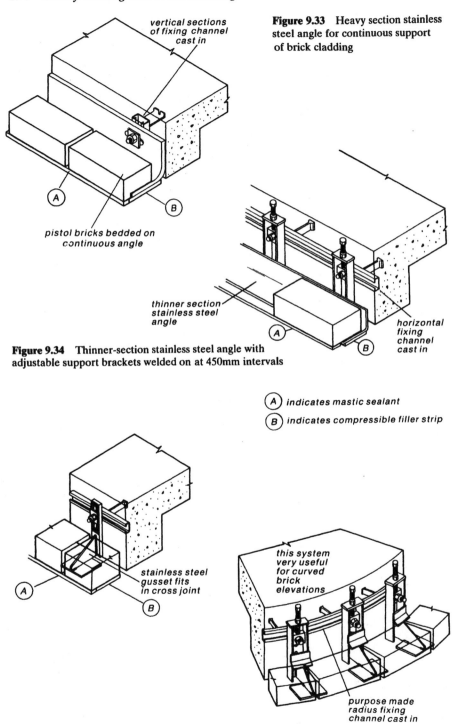

Figure 9.33 Heavy section stainless steel angle for continuous support of brick cladding

vertical sections of fixing channel cast in

pistol bricks bedded on continuous angle

thinner section stainless steel angle

horizontal fixing channel cast in

Figure 9.34 Thinner-section stainless steel angle with adjustable support brackets welded on at 450mm intervals

(A) indicates mastic sealant

(B) indicates compressible filler strip

stainless steel gusset fits in cross joint

this system very useful for curved brick elevations

purpose made radius fixing channel cast in

Figure 9.35 Separate stainless steel brackets supporting each end of stretchers

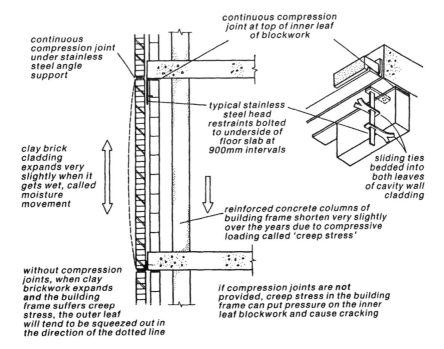

continuous compression
joint at top of inner leaf
of blockwork

continuous
compression joint
under stainless
steel angle
support

typical stainless
steel head
restraints bolted
to underside of
floor slab at
900mm intervals

clay brick
cladding
expands very
slightly when it
gets wet, called
moisture
movement

sliding ties
bedded into
both leaves
of cavity wall
cladding

reinforced concrete columns of
building frame shorten very slightly
over the years due to compressive
loading called 'creep stress'

without compression
joints, when clay
brickwork expands
and the building
frame suffers creep
stress, the outer leaf
will tend to be squeezed out in
the direction of the dotted line

if compression joints are not
provided, creep stress in the building
frame can put pressure on the inner
leaf blockwork and cause cracking

Figure 9.36 The importance of compression joints in brick cladding to framed structures

Pistol bricks

In order to disguise the presence of the supporting steel angle, within the thickness of a mortar bed joint, the first course of bricks may be rebated, as shown in Fig. 9.37. These 'pistol bricks' can be produced as bricks of special shape if the manufacturer is given sufficient notice. Alternatively they can be cut on a masonry bench saw.

Before the importance of horizontal compression joints at every floor level of multi-storey buildings was fully realised, great problems developed from the use of brick 'slips', which were used to disguise concrete toe beams. For these reasons, stainless steel angles or brackets are currently preferred for the support of brick cladding, see Fig. 9.33 to 9.35.

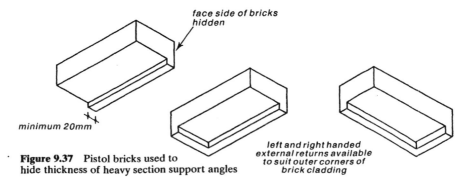

face side of bricks
hidden

minimum 20mm

left and right handed
external returns available
to suit outer corners of
brick cladding

Figure 9.37 Pistol bricks used to
hide thickness of heavy section support angles

Tying back

In addition to providing support at every floor level for panels of external brick cladding, they must be securely tied back to the structural frame of a building. The non-corroding metal ties for this are used in addition to the normal cross-cavity wall ties, required between inner and outer skins of masonry, as shown in Fig. 9.17.

Vertical movement joints

The outer leaf of face brickwork in cavity wall cladding to a framed building will expand and contract due to changes in temperature and moisture content.

The horizontal compression joints shown in Fig. 9.31 to Fig. 9.35 allow up and down movements in the brickwork to take place freely.

Allowance must also be made for sideways expansion and contraction, however, so that the outer leaf of brickwork does not cause damage to itself or the structure.

A typical vertical movement joint, with white expanded plastic foam joint filler projecting, can be seen in Fig. 9.38. This shows 102mm thick brick cladding to a concrete frame building. (Note also the typical stainless steel *permanent* support angle, see Fig. 9.34, for the soldier course of brickwork to be constructed across the window opening. In addition, the *temporary* timber beam, to keep the lower arris of the soldier course perfectly straight at the window head).

Vertical movement joints should be provided at intervals: between 9 and 12m for clay brickwork; not exceeding 7 to 9m for sand lime (calcium silicate) bricks; and at 6m intervals for concrete blockwork.

Plainly, vertical movement joints form a 'straight joint' weakness in a wall, and so must be strengthened with stainless steel slip-ties every fourth course—as shown in Chapter 12 (Fig. 12.10).

Figure 9.38 Brick cladding to reinforced-concrete framed building showing stainless steel angle (Fig. 9.34) to support outer leaf brickwork across window opening

Head restraints

When a building has load bearing walls (see Fig. 9.2), the top of these walls are rigidly held in position by the weight of floors bearing upon them.

The external brick cladding, and internal walls within framed structures, have soft compression joints at the underside of floor slabs, see Fig. 9.31 to 9.35. These must be held in position by some form of head restraint. The inset to Fig. 9.36 shows a typical example of a stainless steel sliding head restraint.

10

Damp prevention

It is important that the bricklayer should understand the causes of dampness and the methods of damp prevention in the construction of a brick wall. Dampness may occur even after precautions have been taken, if those precautions are based on insufficient knowledge resulting in lack of care in the application of damp prevention.

Forming a dpc

The spread of dampness into the main structure of a building is prevented by a damp-proof course (or damp-course). This is a layer of non-absorbent material, examples of which are given in Table 10.1. The damp-proof course must remain intact. All flexible dpc materials must be laid upon a thinly spread bed of fresh mortar, to provide full support across frogs, perforations and joints between bricks. Laying the bed for the next course of bricks ensures that the dpc is then neatly sandwiched and protected. It must not be broken by being laid on an uneven surface, by pebbles in the mortar if a slate damp-course, or by clumsiness on the part of the bricklayer in dropping bricks on to the newly-laid damp-course.

Routes for damp penetration of buildings

Damp affects walls in several ways.
    1. Moisture rising up from the ground.
    2. Penetration through the face of walls, caused by driving rain.
    3. Moisture percolating downwards from the top of walls or chimney stacks.
    In building up any wall construction, and having in mind the causes of dampness, the passage of the moisture should be followed and the necessary precautionary measures should be taken in their logical sequence to prevent it entering the interior of the building see Fig. 10.1.

1. Moisture rising from the ground _____

If soil investigation tests on a site indicate that the ground waters contain soluble sulphates, then the cement used in sub-structure brickwork and foundation concrete must be Sulphate Resisting.
    The construction at the wall base, in conjunction with the floor construc-

Table 10.1 Types of damp proof course

General classification	Description	Type of material	Comments
Flexible	Rolled materials available in a range of widths to suit different wall thicknesses and for folding into cavity trays. Provide the most economical and convenient type of dpc.	Hessian-reinforced bitumen. Ditto with thin lead foil sandwiched within. Fibre reinforced bitumen. Ditto with lead foil.	Standard dpc material available in rolls of approx 8m in length. Must be minimum 100mm running laps, and at corners or T-junctions. Must be laid upon a thin bed of fresh mortar. Always store rolls on end to avoid distortion and cracking. Keep this type of dpc in a warm place before use, to make unrolling easier in cold weather.
		Pitch-polymer. Black polythene.	Thinner, tougher dpc which will not squeeze out under long term pressure from brickwork. Easier to use where projecting vertical dpc must be folded against sides of window and door frames. Available in 24m length rolls. Same minimum laps as above.
	Metal.	Sheet lead. Sheet copper.	Lead and copper dpc should be painted both sides with bituminous paint before use, to prevent possible corrosion from cement, lime or soluble salts in clay bricks. Laps a minimum of 100mm, or joints 'welted', particularly if used to resist downward penetration of dampness. Must be laid upon a thin bed of fresh mortar.
Semi-rigid		Natural mined rock asphalt, or a carefully controlled mixture of bitumen and limestone aggregate.	Heavy blocks of material melted down on site and spread in two or three separate coats while molten. Normally only used as a wall dpc if mastic asphalt tanking to a basement is being applied in the building, (see Figs. 10.14 and 10.15).
Rigid	Engineering bricks, minimum two courses bedded in Group (i) cement mortar 1:3 Only used for horizontal dpc.	Black or red Class 'A' engineering bricks, with a water absorption not exceeding 4.5%	A very good method of creating a dpc in a free standing boundary wall, for reasons given in Chapter 12 and shown in Fig. 12.9. Cross joints may be left 'open', to prevent dampness rising up through the mortar. Engineering bricks are not suitable to resist downwards penetration of dampness.
	Slates, two courses bedded in Group (i) cement mortar 1:3	Sound, hard Welsh slate only. Soft flaky slates are not suitable for a dpc.	A traditional dpc, with the slates bedded 'stretcher bond', which is usually described in Bills of Quantities as 'laid breaking joint' (See Fig. 10.20). Not used in modern construction due to cost, and also because the slightest building settlement will crack the slates. Overall thickness of a slate dpc is 35 to 40mm.

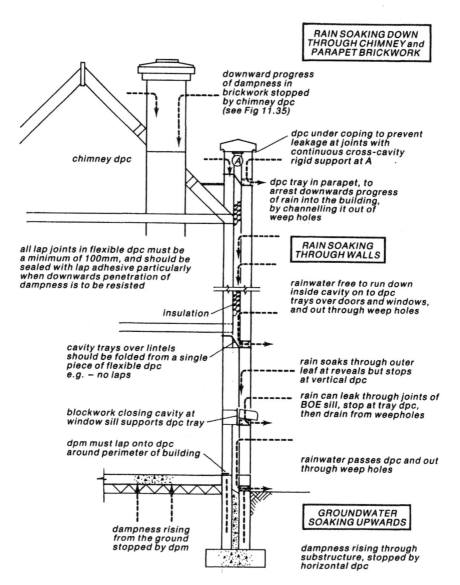

Figure 10.1   Routes for dampness to soak into a building, and dpc provision to resist it

tion, is shown in the following order:
1. timber or suspended floors above ground level;
2. solid concrete floors above ground level;
3. concrete floors below ground (basement floors).

Timber floor above ground level

Location of main horizontal dpc

All timber wall plates, and structural floor joists associated with timber flooring must be pressure impregnated against fungus and beetle attack.

(a) Moisture rises through the foundation concrete and also penetrates the wall in contact with the soil. To prevent it continuing up the wall, some impervious material must be built into the wall; this is called the horizontal dpc, which should be a minimum of 150mm above ground level (Fig. 10.2). A one-brick external wall building has been chosen for the purposes of illustration. (The apprentice should note that this particular wall thickness is unsuitable with regard to weather penetration and thermal insulation, and would therefore require additional constructional treatment to outer and inner faces).

Sealing the ground under a building

(b) Some provision must also be made to prevent damp air rising within the building, carrying with it the odours associated with decaying vegetable and animal matter. This is done by covering the whole of the site within the walls of the building with a layer of concrete from 100mm to 150mm thick; this is called the 'site concrete' or 'over-site' (Fig. 10.3). The over-site must be positioned so that its top surface is not below the highest level of the ground surface or paving around the building.

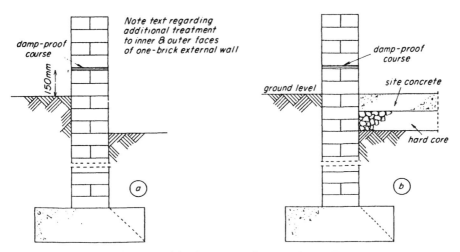

Figure 10.2 Horizontal dpc provision in free standing boundary wall
Figure 10.3   Horizontal dpc provision in external wall of building

Underfloor ventilation against dry rot

(c) Having placed the over-site, the floor must be supported, so as to allow free passage of air—to prevent the floor timbers from rotting. This is achieved by building honeycomb sleeper walls. Two types are illustrated in Fig. 10.6. The sleeper walls nearest to the main wall are positioned as in Fig. 10.4; it provides freedom of movement when the bricklayer is building the wall and less possibility of mortar droppings collecting between the main wall and the sleeper wall. Fig. 10.5 shows a construction that will provide efficient support and very important ventilation to the floor timbers.

Damp, stagnant air provides ideal growing conditions for a fungus commonly called 'dry rot' to take root in constructional timbers.

Although needing moisture to grow, the fungus roots destroy the timber, leaving it split, broken and 'dry' thereby explaining the name.

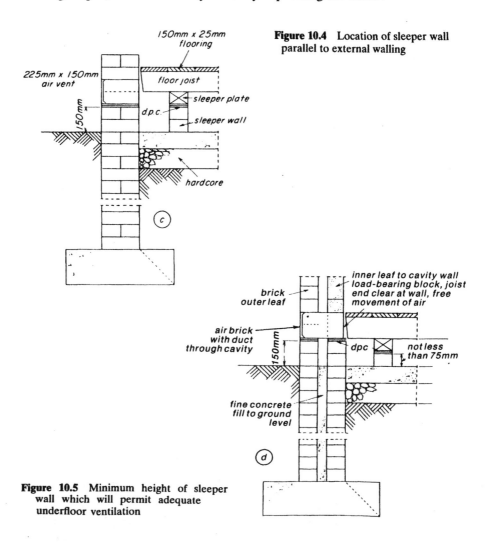

Figure 10.4   Location of sleeper wall parallel to external walling

Figure 10.5   Minimum height of sleeper wall which will permit adequate underfloor ventilation

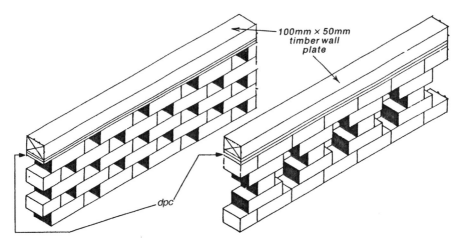

100mm × 50mm
timber wall
plate

dpc

Figure 10.6 Honeycomb sleeper walls

Bedding wall plates

(d) Having taken the necessary precautions against dampness, it is the bricklayer's job to bed the wall plates in readiness for the fixing by the carpenter of joists and flooring. The wall plate, usually of 100 × 50mm or 100 × 75mm sawn timber, is placed immediately above damp-course level. This completes the construction so far as the bricklayer is concerned, with the exception of the building-in of air bricks, which must be placed in suitable positions; their function is to ensure complete underfloor ventilation. It may happen that a hollow timber floor of one room is adjacent to a solid concrete floor in the next room, thus presenting some difficulty with regard to through ventilation. This may be overcome by inserting a series of drain-pipes under the solid floor (Fig. 10.7), thus allowing a satisfactory flow of air to be maintained from front to back with terraced houses particularly.

It will be noted in the illustration that a difference in floor levels occurs. Where the adjacent floor levels are the same, thickening of the solid concrete floor, as shown in Fig. 10.8, will ensure that any joist timbers touching the partition wall are protected from the effects of dampness. This is achieved by similar linking of dpm and dpc. An alternative construction is the application of a waterproof membrane or vertical damp course (Fig. 10.9).

The size of a ground-floor joist in ordinary domestic construction is usually 125 × 50mm laid on edge, with sleeper walls placed at 1350 to 1500mm intervals to give satisfactory support. An alternative type of suspended ground floor construction, using precast concrete beams and blocks, is shown in Chapter 9, Fig. 9.6.

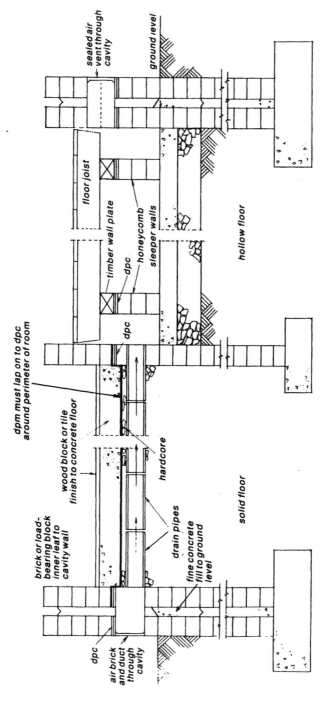

Figure 10.7   Ventilation pipes are laid under the section of solid flooring, to ensure vigorous through-ventilation beneath suspended timber flooring of adjacent rooms, where floor levels are different

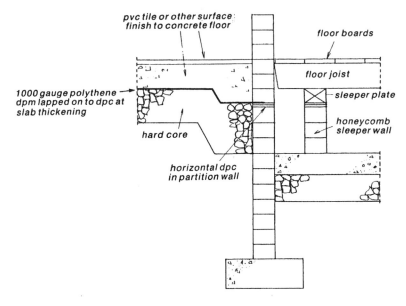

pvc tile or other surface finish to concrete floor

floor boards

floor joist

1000 gauge polythene dpm lapped on to dpc at slab thickening

sleeper plate

honeycomb sleeper wall

hard core

horizontal dpc in partition wall

Figure 10.8 Protecting floor timbers from dampness at the junction of solid and suspended ground flooring

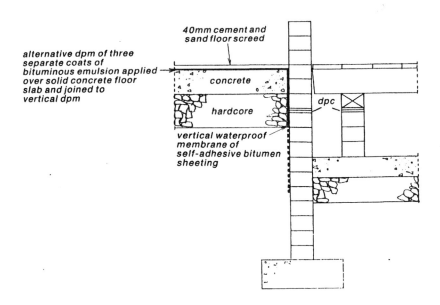

40mm cement and sand floor screed

alternative dpm of three separate coats of bituminous emulsion applied over solid concrete floor slab and joined to vertical dpm

concrete

dpc

hardcore

vertical waterproof membrane of self-adhesive bitumen sheeting

Figure 10.9 Alternative construction to Fig. 10.8

Solid concrete floors above ground level

The sequence of damp penetration can be followed as before. The horizontal damp-proof course level must never be above the floor level. Brick rubble or hardcore laid directly beneath the concrete floor will not only prevent settlement, but, being of a porous nature, will help prevent dampness. Its thickness should be approximately that of the concrete floor (Fig. 10.10). Concrete, used in the floor construction, will not prevent the passage of ground moisture and must therefore be either waterproofed or a waterproof membrane should be incorporated. The latter can be applied either as part of the floor finish, e.g. wood blocks bedded in bitumen (Fig. 10.11), in 'sandwich' layers of concrete (Fig. 10.12) or 1000 gauge polythene placed on a 'blinding coat' of fine quarry waste which is laid immediately above the hardcore (Fig. 10.13), or as three coats of bituminous emulsion, as in Fig. 10.9.

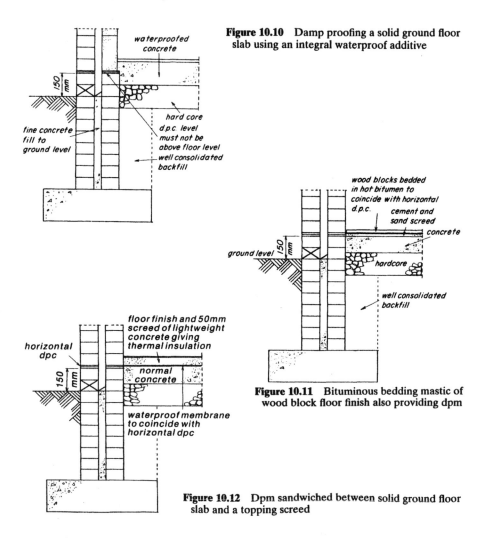

Figure 10.10   Damp proofing a solid ground floor slab using an integral waterproof additive

Figure 10.11   Bituminous bedding mastic of wood block floor finish also providing dpm

Figure 10.12   Dpm sandwiched between solid ground floor slab and a topping screed

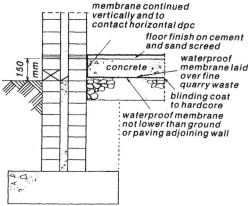

membrane continued
vertically and to
contact horizontal dpc

floor finish on cement
and sand screed

waterproof
membrane laid
over fine
quarry waste

concrete

blinding coat
to hardcore

waterproof membrane
not lower than ground
or paving adjoining wall

150 mm

Figure 10.13 1000 gauge polythene dpm under solid concrete floor

Basement floors below ground level

Moisture penetrates through the walls in contact with the surrounding soil
and upwards through the foundations and the whole floor of the building.
To resist this moisture penetration, the base of a structure must be damp-
proofed throughout the floors and walls below ground level and to at least
150mm above ground level. This is called 'tanking.'

Basement tanking

Mastic asphalt, applied molten in three separate coats, is the only way of
reliably resisting static ground-water pressure. Such as when the general
ground-water level in the surrounding soil is always higher than the
basement floor level.

Alternatively, self-adhesive bituminous sheeting applied to floor and
walls, carefully lapped and sealed at joints, may be considered sufficient for
damp proofing a basement on a well drained site.

Both methods of tanking a basement need permanent protection. A
'loading coat' of *in-situ* reinforced concrete is necessary on basement floors
to 'hold down' and protect mastic asphalt or bituminous sheeting. Protection with brickwork is required to prevent damage, and to hold back
vertical tanking on basement walls.

Tanking from outside

Figure 10.14 shows tanking which has been applied externally. This is the
best method of working, because eventually ground-water pressure will
tend to keep the vertical waterproofing pressed back in place.

The last operation is to build an external, brick wall covering, one half-
brick in thickness, to protect the tanking when the soil is back-filled around
the building.

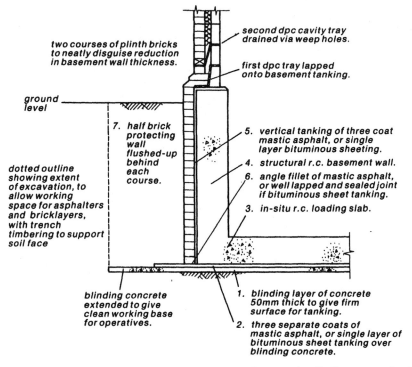

two courses of plinth bricks
to neatly disguise reduction
in basement wall thickness.

second dpc cavity tray
drained via weep holes.

first dpc tray lapped
onto basement tanking.

ground
level

7. half brick
    protecting
    wall
    flushed-up
    behind
    each
    course.

dotted outline
showing extent
of excavation, to
allow working
space for asphalters
and bricklayers,
with trench
timbering to support
soil face

5. vertical tanking of three coat
    mastic asphalt, or single
    layer bituminous sheeting.

4. structural r.c. basement wall.

6. angle fillet of mastic asphalt,
    or well lapped and sealed joint
    if bituminous sheet tanking.

3. in-situ r.c. loading slab.

blinding concrete
extended to give
clean working base
for operatives.

1. blinding layer of concrete
    50mm thick to give firm
    surface for tanking.

2. three separate coats of
    mastic asphalt, or single layer of
    bituminous sheet tanking over
    blinding concrete.

Figure 10.14 Basement tanking. Vertical section through external wall of basement, showing sequence of construction working from *outside*

Tanking from inside

Figure 10.15 shows the opposite way of working, that is tanking applied from inside: the method likely to be used where an *existing* basement is to be waterproofed. The same *in-situ* reinforced concrete floor loading slab, and brick walling protection to vertical tanking are necessary. Care should be taken beforehand to check the reduction in basement headroom that this loading slab thickness will cause.

2. Moisture entering through the face of walls

Some consideration has already been given to this in respect of brick walls below ground level, where vertical damp-proofing is necessary. In walls above ground level exposed to the atmosphere, vertical damp-proofing is in the true sense of the term, unnecessary.

Cavity wall construction, dealt with in detail in Chapter 9, is the standard form of building the external walls of domestic and commercial structures of brickwork and blockwork. The 50 to 75mm wide air space acts as a vertical damp proof 'zone' within all the external walls of a building. The cavity is a complete break between the outer leaf which can become saturated and the internal wall surface.

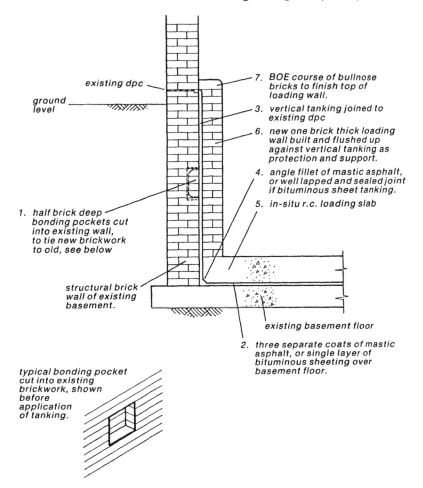

existing dpc

*ground
level*

*7. BOE course of bullnose
bricks to finish top of
loading wall.*

*3. vertical tanking joined to
existing dpc*

*6. new one brick thick loading
wall built and flushed up
against vertical tanking as
protection and support.*

*4. angle fillet of mastic asphalt,
or well lapped and sealed joint
if bituminous sheet tanking.*

5. in-situ r.c. loading slab

*1. half brick deep
bonding pockets cut
into existing wall,
to tie new brickwork
to old, see below*

*structural brick
wall of existing
basement.*

existing basement floor

*2. three separate coats of mastic
asphalt, or single layer of
bituminous sheeting over
basement floor.*

*typical bonding pocket
cut into existing
brickwork, shown
before
application
of tanking.*

Figure 10.15 Basement tanking. Vertical section showing sequence of construction working from *inside*

Solid walls

One brick thick solid walls, if properly built and not exposed to driving rains, will resist the passage of moisture to some degree.

If moisture penetrates a brick wall, it will probably occur where the mortar joint is in contact with the brick (the 'interface'), because of insufficient bond between them. For instance, where a very hard type of mortar is used there is a tendency for the mortar to shrink away from the bricks, leaving hairline cracks through which moisture can pass.

For this reason, one-brick solid walling is unsuitable, together with the fact that it allows heat energy to escape readily by conduction.

Chapter 9 (Fig. 9.4) shows how the insulated cavity space reduces heat loss and provides a barrier to the cross-cavity penetration of dampness.

It should be noted that the solid one-brick thick external walling of existing buildings and maintenance of other brickwork generally, is raked out and repointed, to improve resistance to the weather as well as its stability.

Excessively strong, cement-rich mortar should be avoided when repointing brickwork. A 1:1:5 mix is quite strong enough for the majority of facing bricks, in order to allow the natural weathering processes of wetting and drying, absorption and evaporation to take place evenly, from brick and joint surfaces respectively.

Cavity walls

With those dual requirements of the Building Regulations, that external walls of dwellings shall resist rain penetration and conserve heat energy, standard cavity wall construction as shown in Chapter 9 (Fig. 9.1) is the most economical form of construction for external walling.

Where a cavity is closed at the reveals of windows and doors, a vertical dpc separates inner and outer brick masonry to prevent rain from soaking across.

Where the cavity is 'bridged' by lintels across openings, a 'gutter' or 'tray' of flexible dpc material is formed to gather up any rainwater that has soaked in and direct it out again via weep holes as shown in Fig. 10.1.

Trays of flexible dpc under brick and tile sills similarly prevent rain leakage at joints from progressing to the internal wall surface.

Cavity walling has been examined in general terms, in this present chapter, from a number of directions. Cavity wall constructional details are dealt with in greater depth in Chapter 9.

3. Moisture percolating through the tops of walls _____

Walls below the roof level of a building get wetted by rain on one side only. The brickwork below the overhanging eaves of a pitched roof, as example shown in Chapter 9 (Fig. 9.2), is very nicely protected at the top.

Parapet walls

These are brick walls that stand up around the edges of a roof. They provide somewhere against which to seal the roof covering of asphalt or sheet material, and can also act as a permanent barrier around a roof.

However, the brickwork of a parapet wall, as illustrated in Fig. 10.1 and Figs. 10.16 and 10.17 is exposed to the weather on both sides and the top. For this reason great care must be taken to prevent dampness entering the walls of buildings in this way.

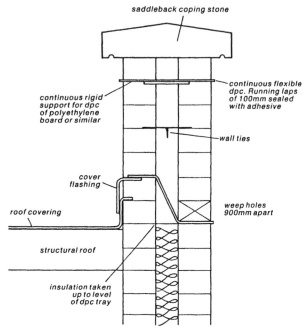

saddleback coping stone

continuous rigid
support for dpc
of polyethylene
board or similar

continuous flexible
dpc. Running laps
of 100mm sealed
with adhesive

wall ties

cover
flashing

roof covering

weep holes
900mm apart

structural roof

insulation taken
up to level
of dpc tray

Figure 10.16   Parapet cavity tray tilted away from roof

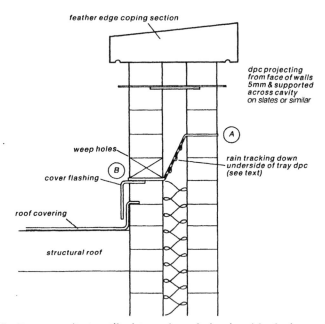

feather edge coping section

dpc projecting
from face of walls
5mm & supported
across cavity
on slates or similar

A

weep holes

B

rain tracking down
underside of tray dpc
(see text)

cover flashing

roof covering

structural roof

Figure 10.17   Parapet cavity tray tilted towards roof, showing risk of rain penetration at A
and B, channeling rain on to the inner leaf in areas of high exposure

Chimney stacks

This brickwork is open to the elements on *four* sides as well as the top, and also suffers the stress of thermal expansion and contraction due to the passage of hot gases within.

For these reasons, it is not surprising that chimney stack brickwork is generally the first that needs repointing or some other maintenance after only twenty years into the life of the building. Choice or specification of bricks and mortar are therefore particularly important for chimney stack brickwork.

Choice of bricks

Facing bricks in the exposed locations of parapet walling and chimney stacks, need to be resistant to frost. This requirement obviously influences the choice of facing bricks for the whole building, because parapet and chimney stack brickwork must match colour and type of the rest of the facework.

Choice of mortar

Reference should be made to Chapter 2 (Table 2.8) for a mortar strength also suited to these exposed positions. The use of Sulphate Resisting cement in the bricklaying mortar for parapets and chimneys may be advisable if the clay bricks to be used contain soluble salts. This to avoid long term deterioration of this exposed brickwork by sulphate attack if Ordinary Portland cement were to be used.

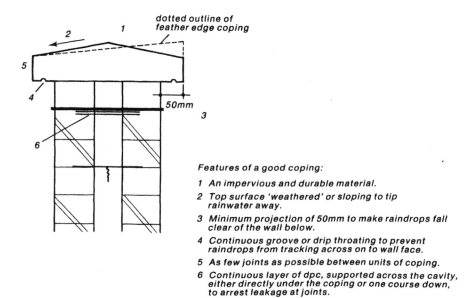

Features of a good coping:

1 An impervious and durable material.
2 Top surface 'weathered' or sloping to tip rainwater away.
3 Minimum projection of 50mm to make raindrops fall clear of the wall below.
4 Continuous groove or drip throating to prevent raindrops from tracking across on to wall face.
5 As few joints as possible between units of coping.
6 Continuous layer of dpc, supported across the cavity, either directly under the coping or one course down, to arrest leakage at joints.

Figure 10.18   Section through parapet wall with saddleback coping

Weathering to the top of walling

Some form of *projecting* coping is preferable to protect the top of walls, to function rather like an umbrella.

The features of a good coping for a cavity wall parapet around a building, shown in Fig. 10.18, are similar to those detailed in Chapter 12 for solid boundary walling.

Precast concrete or natural stone copings with the features illustrated in Fig. 10.18 provide the best long term protection to parapet brickwork.

Similarly the type of chimney stack coping illustrated on Fig. 10.1 and also in Chapter 11 (Fig. 11.35) include the foregoing desirable features to ensure efficient long term performance.

The variety of BOE cappings shown in Fig. 10.19 look more attractive, but have many more joints, and for this reason are less successful in preventing rainwater from saturating parapet brickwork and the tops of other free standing walls.

A number of brick manufacturers have developed two-part brick systems of cappings and copings for walls, to improve the performance of the parapet dpc plus their solidity and appearance. See some examples Chapter 12: Fig. 12.15.

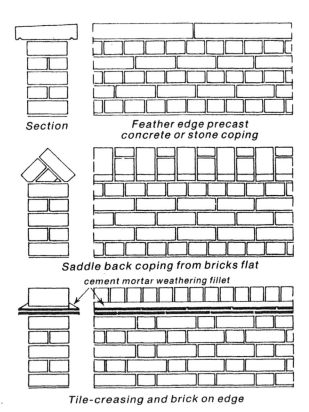

Section   *Feather edge precast concrete or stone coping*

Saddle back coping from bricks flat

cement mortar weathering fillet

Tile-creasing and brick on edge

Figure 10.19   Finishes to tops of walls

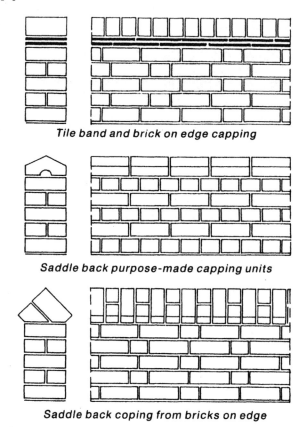

Tile band and brick on edge capping

Saddle back purpose-made capping units

Saddle back coping from bricks on edge

Figure 10.19 *continued*

Parapet wall construction

Many parapet walls have been built as one brick or $1\frac{1}{2}$ brick thick solid walling, but it has been found that cavity wall construction for parapets is preferable in preventing downwards penetration of dampness.

A dpc tray incorporated across the cavity wall width collects any rainwater that enters the cavity from either direction and disperses it via weep holes, see Fig. 10.16.

Purpose made nylon fibre plugs can be put into these weep holes to prevent any water staining of the facework.

Tilting the cavity tray the opposite way, towards the roof, as shown in Fig. 10.17 runs the risk of directing dampness on to the inner leaf of masonry.

Types of damp-proof course _____

Materials selected for damp-proof courses must be capable of resisting the passage of moisture, and when applied to the wall must be continuous throughout. This continuity may be broken either by bad workmanship or, where a rigid type of damp-proof course is used, e.g. slate, by settlement of the building. Failures will therefore be prevented by the selection of a suitable material to supplement good craftsmanship and a reliable foundation.

Slates

See Table 10.1 (page 199) and Fig. 10.20.

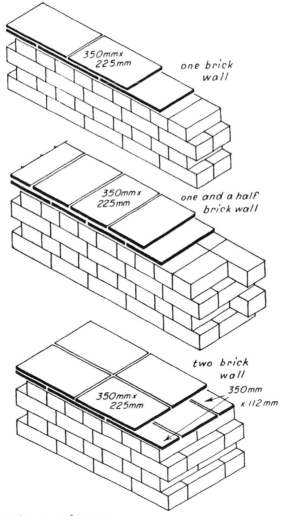

Figure 10.20 Slate damp proof courses

11

Fireplace construction

Many centuries ago when people first decided to bring the fire they used for cooking and heating inside their homes, the safest place for it was in the middle of the ground floor where everyone could keep an eye on it.

Smoke simply rose up into the rafters to escape through spaces in the roof covering.

Not until houses were built from non-combustible stone and brick was the fire placed against the wall between two attached piers called 'jambs' to form a fireplace recess.

Eventually the two jambs were joined with an arch and a chimney breast constructed above, leading to a hole in the wall behind for the smoke to escape.

Later, a vertical shaft was formed in the chimney breast to channel smoke up through the roof, inside a chimney stack of stone or brick, in order to protect the roof timbers from fire risk.

In order that the apprentice may clearly understand this important craft operation, this chapter has been divided into three main headings:

1. The types and positioning of fireplaces, flues, and chimneys, as desired by the architect.

2. The Building Regulations controlling construction.

3. The construction of a fireplace through the ground floor, first floor, and roof of an ordinary domestic dwelling. By following the sequence of operations, the reader should have no difficulty when confronted with a problem.

1 Types of fireplace

Single fireplaces The fireplace opening is formed on one side of the wall only, by the formation of attached piers called 'jambs.' Figures 11.1 and 11.2 illustrate the plans of various jamb arrangements at the ground floor level or wherever the base or a chimney breast is commenced. Fig. 11.1(a) shows a simple arrangement where the main wall forms the back of the fireplace opening. Fig. 11.1(b) illustrates a similar example where the main wall is 1½ bricks thick and the back of the fireplace is 1 brick thick. Figs. 11.1(c) and 11.2 show typical examples of fireplaces placed on external walls; the projection of the jambs into the room is lessened by the formation of an external breast or the breaking of the wall line on the

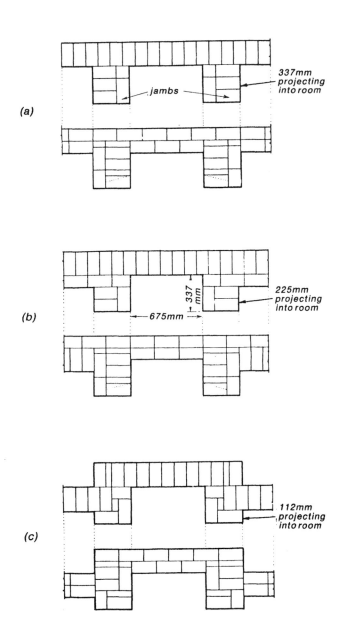

Figure 11.1 Single fireplaces, English bond

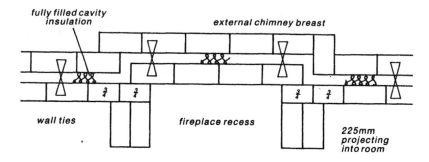

Figure 11.2   Single fireplace on external cavity wall

opposite side of the wall. The advantage gained is the larger area available in the room in which the fireplace is situated.

The width and height of the fireplace opening depends on the type of stove or grate to be inserted, while the width of the jambs depends upon the width of the chimney breast required on the upper floors. See also the positioning or grouping of chimneys and fireplaces (page 223).

The minimum depth of the fireplace opening is 338mm, to allow a 225mm flue to be formed together with a covering of 112.5mm of brickwork when the chimney breast is being constructed above the lintel or arch level of the fireplace opening.

The plans of the alternate courses of bonding are shown for each arrangement; the three-quarter bat on the outside edge of each jamb should be noted. These jambs eventually form the main corners of the chimney breast, and an exception to the bonding rules therefore occurs.

Double or back-to-back fireplaces (Fig. 11.3).
This is a typical example of fireplace construction in the terrace or semi-detached type of dwelling. The fireplaces are formed on the party wall. Fig. 11.4 shows a method of bonding adopted by some bricklayers to save the labour of cutting. This practice should be discouraged as bricks are often omitted where the straight joints occur and 'pockets' are formed, which are filled with rubbish—an example of bad workmanship.

Interlacing fireplaces.
These are fireplaces constructed on an internal wall and placed side-by-side; this arrangement lengthens the

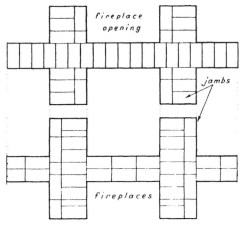

Figure 11.3   Back to back fireplaces

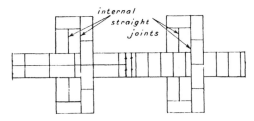

Figure 11.4   An example of bad bonding practice

chimney breast, but the projection of the jambs into the room is reduced, giving greater room area (Fig. 11.5).

Angle fireplaces.   Fig. 11.6 shows the plans of the alternate courses of brickwork bonding at ground level. This is a difficult type of fireplace to construct because it entails a number of twists of the flue which are necessary to obtain correct positioning. Its construction will therefore not be shown at this stage.

Upper floor fireplaces.   Figures 11.7 to 11.11 illustrate the bonding arrangements of upper floor fireplaces in the various types of construction. The alternate arrangement of flues in the interlacing fireplaces should be noted; this will be better understood when the grouping of fireplaces and chimneys has been considered.

Grouping of fireplaces and chimneys

In order to achieve sound construction and to reduce the number of chimney stacks emerging through the roof, fireplaces and chimneys must be grouped to a central position. The fireplaces of each successive floor should therefore be positioned one above the other, and where possible,

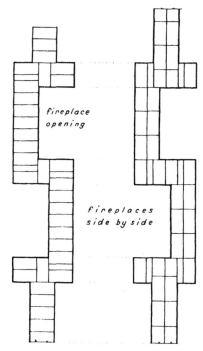

fireplace
opening

fireplaces
side by side

Figure 11.5   Interlacing fireplaces, ground floor

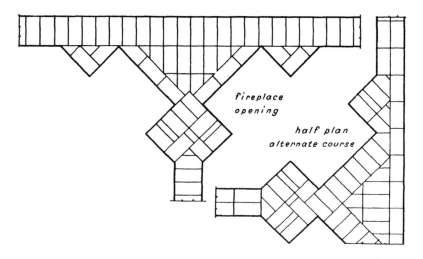

fireplace
opening

half plan
alternate course

Figure 11.6   Plan of angle fireplace, English bond

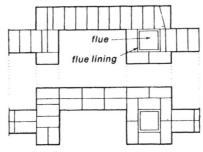

Figure 11.7 Single fireplaces, first floor

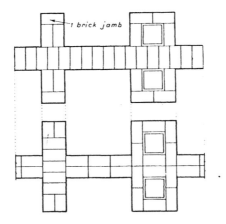

Figure 11.8 Back to back fireplaces, first floor

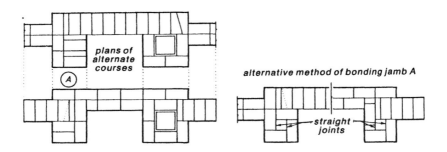

Figure 11.9 Single fireplace at first floor showing alternative methods of bonding wider left hand jamb

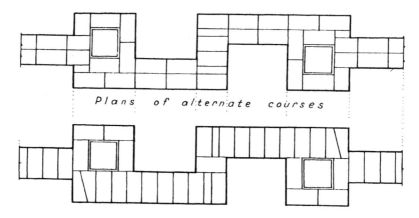

Figure 11.10 Interlacing fireplaces, upper floors

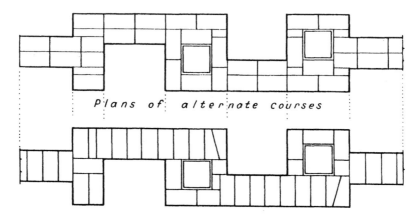

Figure 11.11 Interlacing fireplaces, upper floors, with central flue position

the flues from fireplaces in adjacent rooms will be gathered together before emerging through the roof. The base of a chimney breast must therefore be wide enough to support any type of fireplace construction that occurs in the floors above. Where buildings extend to four or five floors this is not always practicable, and in this case it is permissible to extend the upper floor chimney breasts in their length by corbelling. In simple domestic construction the latter point does not arise.

Figure 11.12 shows single breast fireplace construction on an external wall. The passage of the flues has been marked by dotted lines, and it will be noted that the flue from the ground floor fireplace has been gathered over to miss the first floor fireplace, while at roof level the flues from adjacent rooms have been grouped to form a single stack of four flues. The external chimney breasts have been connected by a face-brickwork semi-circular arch. The alternative is to continue the single chimney breast and to reduce it to stack size by means of circular ramps or by tumbling (Fig. 11.13), where it will develop into a two-flued stack.

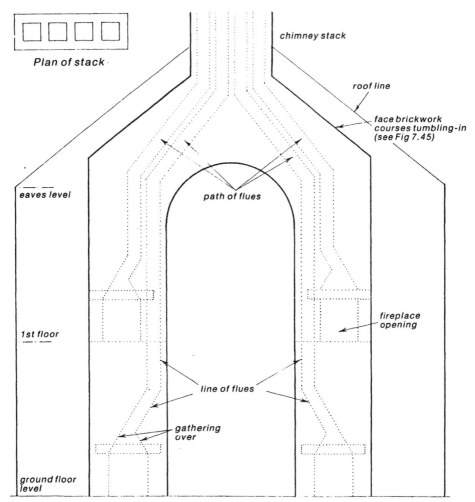

Plan of stack

chimney stack

roof line

face brickwork
courses tumbling-in
(see Fig 7.45)

eaves level

path of flues

fireplace
opening

1st floor

line of flues

gathering
over

ground floor
level

Elevation of outside

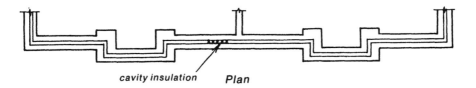

cavity insulation   Plan

Figure 11.12   Single fireplaces on an external wall

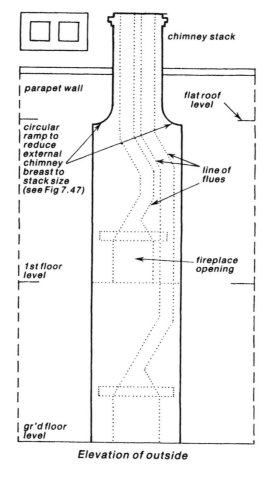

chimney stack

parapet wall

flat roof
level

circular
ramp to
reduce
external
chimney
breast to
stack size
(see Fig 7.47)

line of
flues

1st floor
level

fireplace
opening

gr'd floor
level

Elevation of outside

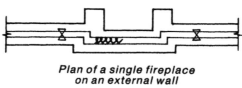

Plan of a single fireplace
on an external wall

Figure 11.13 External chimney breast reduced to stack width using brick ramps (see Fig. 7.46)

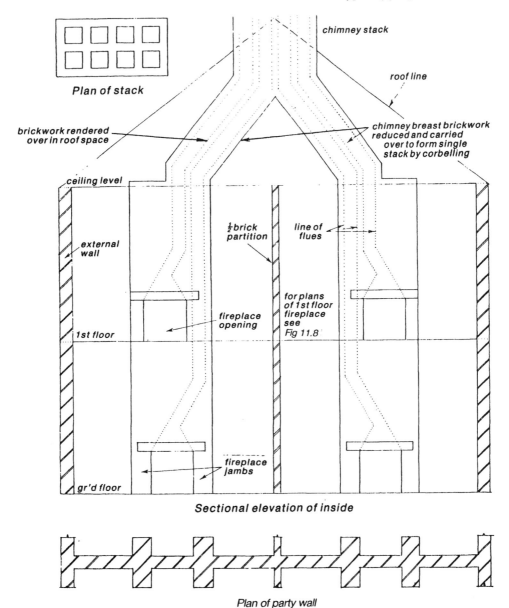

Plan of stack

chimney stack

roof line

brickwork rendered over in roof space

chimney breast brickwork reduced and carried over to form single stack by corbelling

ceiling level

external wall

½ brick partition

line of flues

for plans of 1st floor fireplace see Fig 11.8

1st floor

fireplace opening

fireplace jambs

gr'd floor

Sectional elevation of inside

Plan of party wall

Figure 11.14 Back to back fireplaces on a party wall

Figure 11.14 shows the grouping of chimneys and fireplaces on a double-breast party wall. The grouping of fireplaces from adjacent rooms takes place within the roof space and does not require the decorative finish as in the external chimney breast, illustrated in Fig. 11.13, which accounts for the set-off a few courses above first floor ceiling level. The chimney stack emerging from the centre of the roof consists of eight flues, but the chimney breasts can be carried up individually, thus emerging one on each side of the ridge of the roof as a four-flue stack.

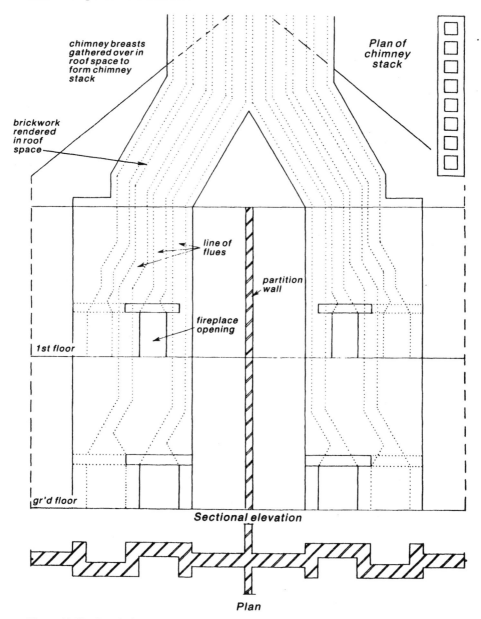

chimney breasts
gathered over in
roof space to
form chimney
stack

Plan of
chimney
stack

brickwork
rendered
in roof
space

line of
flues

partition
wall

fireplace
opening

1st floor

gr'd floor

Sectional elevation

Plan

Figure 11.15   Interlacing fireplaces on a party wall

Figure 11.15 shows the grouping of interlacing fireplaces. In the description of types of fireplaces two methods of arranging the upper floor
fireplaces were shown; both have their merits. Fig. 11.11 is considered to be
more straightforward, while Fig. 11.10 gives symmetrical planning. The
chimney-stack consists of eight flues in a line.

2 Regulations controlling the construction of fireplaces and chimneys _____

The Building Regulations are concerned with stability, fire hazards and the discharge of products of combustion to the outside air. For the purpose of relating text and diagrams, sections will be numbered from No. 1 onwards, so that they will differ from references in the Building Regulations.

These cover the construction of flues, chimneys, fireplace recesses and hearths, for the purpose of an ordinary domestic fireplace appliance which will not exceed an output rating of 45 kW and a resulting flue not exceeding approximately 225mm × 225mm. Other clauses modify construction where gas appliances are used. The following descriptions are therefore an interpretation of the regulations in this sense, and this knowledge is sufficient for the construction of ordinary domestic fireplaces.

A chimney may be defined as the solid material surrounding a flue, while the flue is a duct through which smoke and other products of combustion pass. A chimney is therefore a single flue surrounded by at least 100mm of brickwork and a stack consists of two or more flues surrounded by the requisite amount of brickwork. If the 'withes' or 'midfeathers' (dividing walls between flues) of a stack are $\frac{1}{2}$ brick in thickness (Fig. 11.30, page 241) this will meet the requirements.

A flue formed in a brick wall is normally constructed to the dimensions of 225mm × 225mm or one brick by one brick; the cross-sectional area is *reduced* by the application of purpose-made flue liners. The minimum size of the flue to suit a 45 kW closed appliance is 175mm diameter.

In simple residential buildings the size of a flue used solely for the purpose of discharging the fumes of a gas appliance into the open air must not be less than $12\,000$ mm^2 in cross-sectional area if circular, and $16\,500$mm^2 if rectangular. Proprietory purpose-made flue blocks are available for the construction of gas flues, which must have a minimum dimension on plan of 90mm.

1. Every flue must be surrounded by at least 100mm thickness of brickwork, properly bonded and exclusive of the thickness of the flue lining.

2. Every chimney or stack must be built to a height of at least one metre above the last point of contact where it emerges from the roof. In the case of a gas flue this height is modified, but in no case must a chimney or stack be built up to a height greater than $4\frac{1}{2}$ times its least width at the last point of leaving the roof, unless it is adequately supported against overturning (Figs. 11.16 and 11.17). The Building Regulations state that the top of a chimney carried up through the ridge of a roof, or within 1000mm of it, and which has a slope on both sides of not less than ten degrees with the horizontal must be at least 600mm above the ridge, but in all other cases at least 1 metre measured from the highest point in the line of junction with the roof.

3. Every chimney must be built on suitable foundations approved by the local Building Control Officer. In many traditional type houses chimney breasts have been projected from the main wall at an upper floor level; Fig. 11.18 illustrates a possible case in which the chimney breast of the

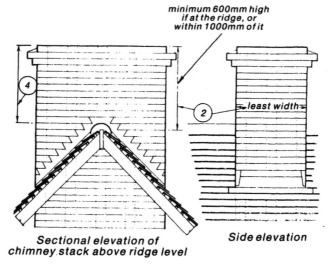

minimum 600mm high
if at the ridge, or
within 1000mm of it

least width

Sectional elevation of
chimney stack above ridge level

Side elevation

Figure 11.16   Lead flashings to weatherproof junction between roof covering and a chimney stack that emerges through the ridge

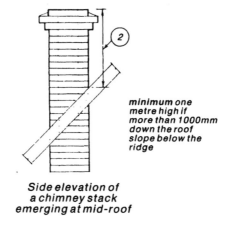

minimum one
metre high if
more than 1000mm
down the roof
slope below the
ridge

Side elevation of
a chimney stack
emerging at mid-roof

Figure 11.17   One metre minimum height where chimney stack emerges mid roof

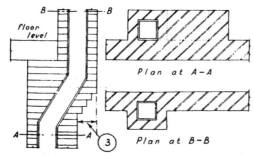

floor
level

Plan at A–A

Plan at B–B

Figure 11.18   Corbelling at first floor level to form an external chimney breast

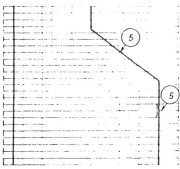

Elevation of chimney breast tumbled-in

Figure 11.19 Travelling a flue in an external chimney

ground floor is discontinued on the upper floor and in consequence the chimney has been corbelled-out on the external wall.

4. The twelve highest courses of every chimney or stack should be built in cement mortar (Fig. 11.16). A chimney stack passing through a roof must be properly protected against downward penetration of dampness. This is achieved by building in a damp proof course within the brickwork, see Figure 11.35.

5. Where chimney flues are inclined, the 'angle of travel' shall not be less than 60° to the horizontal. Chimney flues must be provided with a means of inspection and cleaning, via a gastight door set in a metal frame, see Figure 11.9.

6. The inside of every chimney must be lined with non-combustible materials, e.g. fireclay or terracotta flue liners (Fig. 11.28); this work being carried out as the building of the chimney proceeds. The non-combustible linings are purpose-made units either square or circular in section to fit a one brick by one brick flue.

Flue liners must always be placed with sockets or rebates uppermost, to prevent leakage of any condensation which might form on the inside surfaces. Flue liners must also be jointed with fireproof mortar.

7. It is advisable that the outer surface of every chimney that is within a building or roof space is rendered up to the level of the outer surface of the roof or gutter. (Plastering to the internal walls of rooms largely takes care of this.) This rendering is for the additional protection of any combustible materials adjacent to the chimney where it passes through floors and roof.

(a) Woodwork must not be placed under any fireplace opening within 250mm from the upper surface of the hearth except for the fillets or bearers supporting the hearth, or if there is an air space of not less than 50mm between the underside of the hearth and any combustible material (see Fig. 11.26 and 11.27).

(b) Woodwork must not be built into any wall nearer than 200mm measured to the inside surface of a flue or to the inside of a fireplace recess. (Fig. 11.20).

(c) Woodwork must not be placed closer than 40mm to the surface of a chimney or fireplace recess, excepting floor boarding and skirting.

(d) If the surround of a fireplace opening is constructed of wood it should

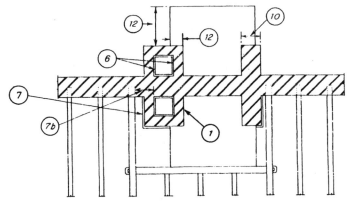

Figure 11.20 Plan of back to back fireplaces at first floor level (for numerical references, see text)

be at a distance of at least 150mm measured horizontally and 300mm measured vertically from the fireplace opening, and it must be backed with solid non-combustible material.

8. Where a chimney is adjacent to constructional steelwork or reinforced concrete, it is advisable to take precautions to ensure that the steelwork etc., is not affected by heat, with metal fixings at least 50mm away.

9. Each fireplace must have its own flue taken to the outside air, with no branching of more than a single fireplace per flue.

10. The jambs of every fireplace opening must be at least 200mm thick (Fig. 11.20).

11. The back of every fireplace opening on a party wall or party structure must be at least 200mm thick.

This 200mm thickness of brick masonry is also necessary to separate flues on either side of a party wall up to the level of the top floor ceiling. Reduction to 100mm thickness between flues takes place at this point, where the bulk of the chimney breast is discontinued and the flues are grouped into one stack for passing through the roof covering (see Figure 11.26, page 236).

It is desirable to keep the air in a flue warmer than the outside atmosphere, so that the air in a flue is rising continuously; the warmer the air in a flue, the quicker is the flow of air, and with it the gases of combustion, while the colder the air, the slower will be the flow, a condition which can lead to 'downdraught' and smoky chimneys.

12. The constructional concrete hearth must extend at least 150mm beyond each side of the fireplace opening and must project at least 500mm in front of it. (Fig. 11.20).

13. The hearth must be at least 125mm thick, formed of incombustible materials.

3 The construction of a fireplace through the ground floor, first floor, and roof ___

It should be noted that solid fuel appliances are frequently omitted from upper floor levels and construction is therefore simplified (Fig. 11.21).

However, it is felt that the apprentice should be aware of traditional practice and of the problems that might be encountered in refurbishment of buildings. A back-to-back fireplace on a party wall has therefore been selected for the purpose of description, as this should cover adequately all types of fireplace construction.

Figure 11.22 illustrates the plan, sectional elevation, and section of ground floor fireplace construction.

The chimney breast is 1.349 m in width, made up of two 330mm jambs and a 675mm fireplace opening, this being sufficiently wide to contain the upper floor fireplace consisting of a 225mm jamb, a 450mm chimney jamb containing the flue, and a 675mm fireplace opening. See also Fig. 11.23 set in a solid ground floor.

Fender wall

A one-brick fender wall supports the constructional hearth and is built up on the site concrete. It is advisable to make this wall one-brick in thickness, rather than a half-brick, as it serves the double purpose of supporting the hearth and part of the timber floor. The void beneath the hearth should be filled with clean hardcore consisting, if possible, of broken brick.

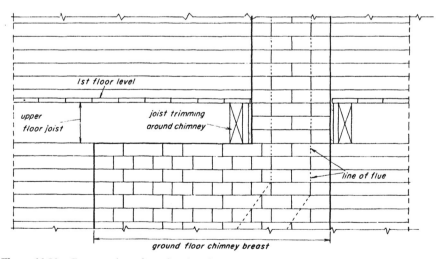

Figure 11.21 Construction where fireplace is not required at first floor level

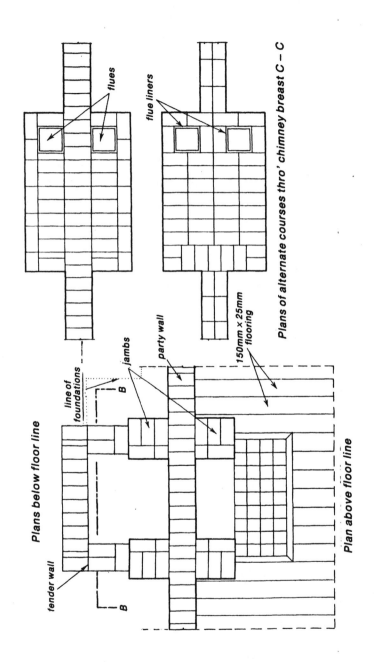

flues

flue liners

Plans of alternate courses thro' chimney breast C – C

line of foundations

jambs

party wall

150mm × 25mm flooring

B

Plans below floor line

fender wall

B

Plan above floor line

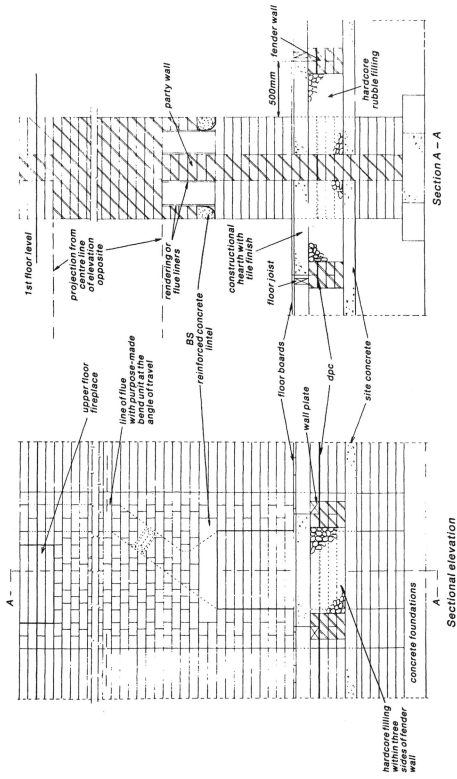

Figure 11.22 Ground floor fireplace set in suspended timber flooring

Section A – A

1st floor level

500mm fender wall

hardcore rubble filling

party wall

projection from centre line of elevation opposite

rendering or flue liners

constructional hearth with tile finish

floor joist

BS reinforced concrete lintel

floor boards

wall plate

dpc

site concrete

Sectional elevation

upper floor fireplace

line of flue with purpose-made bend unit at the angle of travel

hardcore filling within three sides of fender wall

concrete foundations

A —

A —

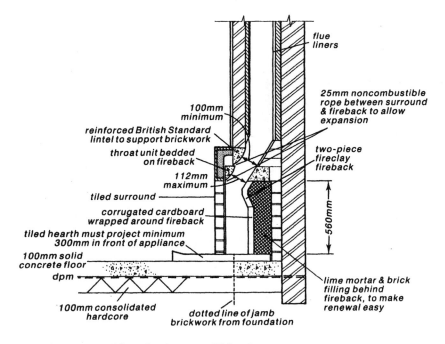

Figure 11.23   Ground floor fireplace on solid flooring

The ground floor of many properties has an *in-situ* cast solid concrete floor or suspended flooring of pre-stressed concrete beams with 440 × 215 × 100 blocks between.

In such cases, there is no fender wall, and the whole floor is of course non-combustible, not just the minimum requirement of 500mm in front of any fireplace.

Flue liners

The path of the flue has been shown by dotted lines. Purpose made bends of fireclay or terra cotta are used to provide a continuous smooth inside surface, (see Fig. 11.28), allowing a flue to be 'travelled' to left or right within the chimney breast brickwork.

Flue liners are continuously bedded and jointed ahead of the courses, and surrounding brickwork is cut and solidly bedded around them as shown in Figs. 11.24 and 11.25.

Great care must be taken to see that flue liners are always set and bedded with the rebate or socket *uppermost*. This is to ensure that any acidic condensation which forms inside a cold flue, when a fire is lit, cannot seep out of the joints between flue liners and attack the surrounding brickwork. Fig. 11.24 is an isometric detail of the gathering over. Note the arrangement of bricks to give quarter bond and a complete tying-in of the cut bricks.

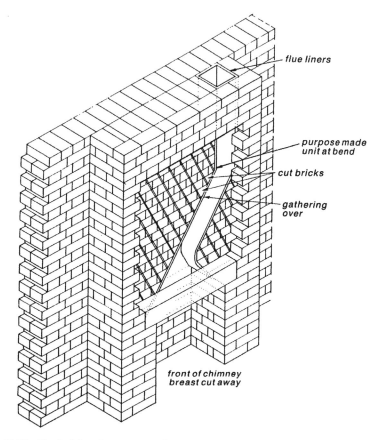

flue liners

purpose made
unit at bend

cut bricks

gathering
over

front of chimney
breast cut away

Figure 11.24   Typical fireplace construction

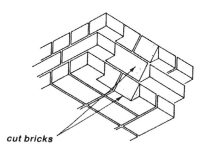

cut bricks

Figure 11.25   Gathering over brickwork around flue linings

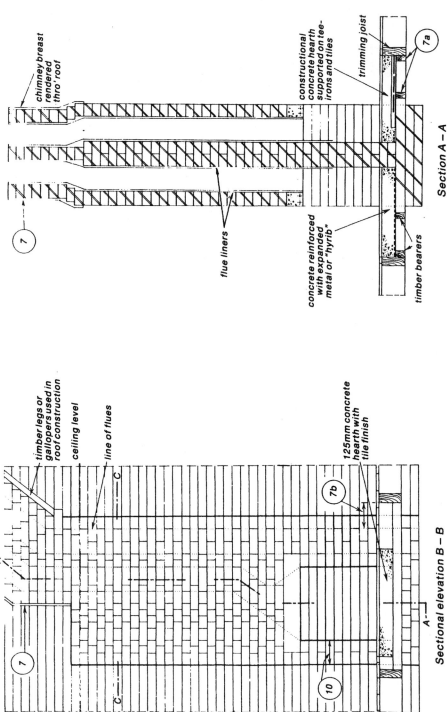

Figure 11.26   Upper floor back to back fireplace construction in suspended timber flooring

The use of flue liners has overcome the problems associated with the formation of a smooth flue, using internal rendering (or parging in older buildings), before purpose-made flue liners were required by the Building Regulations after 1965.

Figure 11.26 illustrates the construction of an upper floor fireplace. It will be seen that the section has been taken through the centre of the fireplace and continued throughout one of the flues. This enables the construction immediately above ceiling level to be clearly illustrated, and it is also amplified in the detail (Fig. 11.27). The chimney breast has been set off to the proper stack size and can either be carried straight up or grouped with the adjacent chimney breast to form a single stack as shown in Fig. 11.29.

Two methods are shown for the construction of the upper floor concrete hearth; one is reinforced with expanded metal strengthened with pressed steel tees, such as 'hyrib,' and the other is formed by a series of short tee-irons placed at convenient centres to receive roofing tiles (with nibs removed), or creasing tiles, which in turn support the concrete. Several methods of permanent non-combustible support are possible, but in every case the concrete must be adequately reinforced.

Figure 11.29 illustrates the chimney construction through the roof. The chimney breasts, reducing to proper size, are grouped to form a single stack. This is achieved by the use of temporary timber legs or 'gallopers,' supported by a corbel projected from the chimney breast as illustrated. The section has been taken throughout one of the flues. The angled 'travel' of these flues, corbelled out from the party wall, is controlled by the purpose-made flue liner units used at the bends. Temporary timber gallopers should be erected as soon as the corbel has been sufficiently tailed-down by the brickwork immediately above and has the stability to bear the weight of the gallopers.

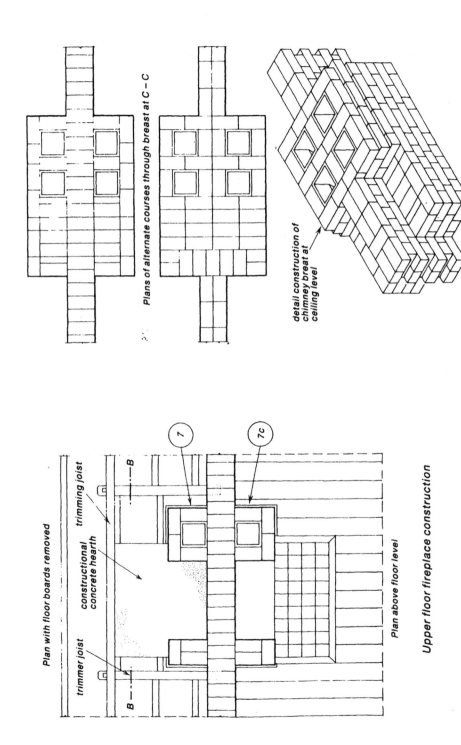

Plans of alternate courses through breast at C – C

detail construction of chimney breat at ceiling level

Plan with floor boards removed

trimming joist

constructional concrete hearth

trimmer joist

Upper floor fireplace construction

Figure 11.27 Reduction of back to back chimney breast to chimney stack size at ceiling level

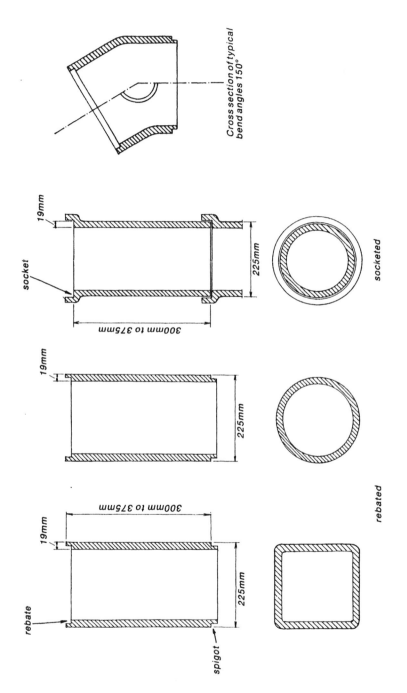

Figure 11.28 Square and circular section flue liners

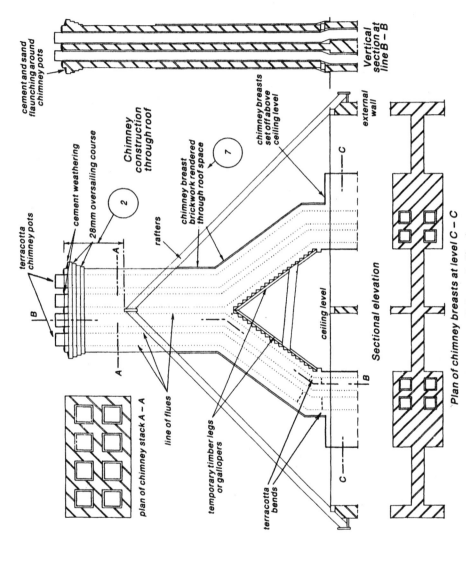

cement and sand flaunching around chimney pots

Vertical section at line B – B

terracotta chimney pots

cement weathering

28mm oversailing course

Chimney construction through roof

chimney breast brickwork rendered through roof space

chimney breasts set off above ceiling level

external wall

rafters

B

A

A

B

plan of chimney stack A – A

line of flues

temporary timber legs or gallopers

terracotta bends

ceiling level

Sectional elevation

C

C

B

Plan of chimney breasts at level C – C

Figure 11.29 Gathering separate flues on either side of a party wall, by corbelling into a single chimney stack, to penetrate the roof at one point only

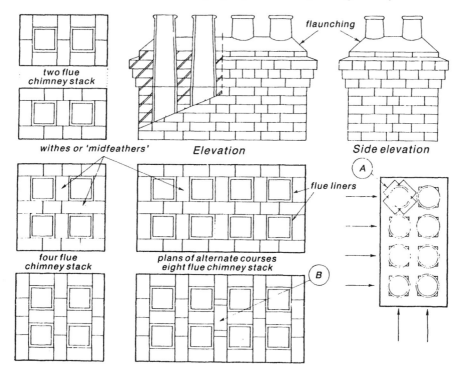

Figure 11.30   Bonding of two, four and eight flue chimney stack

Chimney stacks

Figure 11.30 shows the plans of alternate courses of various sized stacks, and a sectional elevation illustrating the bedding of the chimney-pots. It is advisable to place pieces of slate directly under the pots to assist in their correct alignment and to prevent mortar from dropping down the flues where these are not completely covered by the chimney-pot (A). As the pots are usually uneven in shape, it is impracticable to line them in with a straight-edge, but they should be lined in by the eye in the direction of the arrows. This may appear superfluous, but as a line of chimney-pots can easily be seen from ground level, any lack of alignment presents an unsightly appearance and is indicative of careless workmanship.

B is an alternative method of bonding the 'withe' walls; it is used on the internal withe walls and makes use of the closers that would otherwise be wasted.

Figures 11.31 and 11.32 shows two examples of finish to the top of a stack. There are many excellent examples to be seen in all parts of the country where the stacks have been carefully designed by the architect as a special feature of the building.

In Fig. 11.32 the outer walls of the stack are built in Flemish bond, 1-brick thick, with inner walls or 'withes' of $\frac{1}{2}$-brick. The outer walls are more than $\frac{1}{2}$-brick in thickness, so that excessively cold flues may to some extent be avoided.

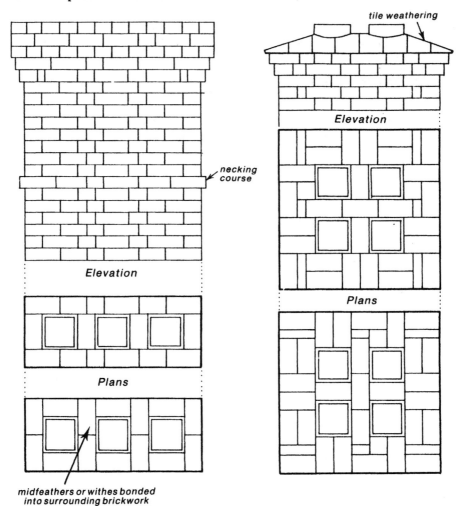

necking
course

midfeathers or withes bonded
into surrounding brickwork

Figure 11.31 Bonding of three-flue chimney stack

Figure 11.32 Bonding of four-flue stack with 1-brick thick outer walls and $\frac{1}{2}$ brick midfeathers or withes. 225mm thick outer brickwork for additional strength and wind resistance, where a chimney stack must be raised higher above the surrounding roof. (Note: the maximum unsupported height for a chimney stack is $4\frac{1}{2}$ times its least dimension on plan

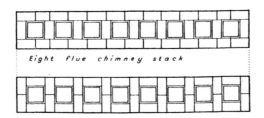

Eight flue chimney stack

Figure 11.33 Bonding chimney stack of eight flues in line

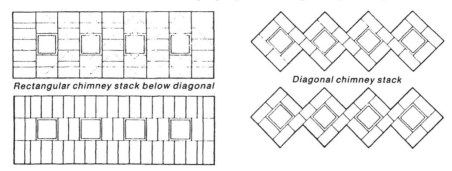

Rectangular chimney stack below diagonal

Diagonal chimney stack

Figure 11.34 Bonding of diagonal chimney stack

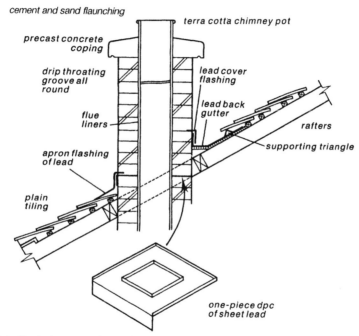

cement and sand flaunching

precast concrete coping

terra cotta chimney pot

drip throating groove all round

lead cover flashing

lead back gutter

flue liners

rafters

apron flashing of lead

supporting triangle

plain tiling

one-piece dpc of sheet lead

Figure 11.35 Typical construction and waterproofing at roof level

Figure 11.33 shows the plans of the alternate courses of bonding in an eight-flue stack, with the flues in line.

Figure 11.34 illustrates the planning of a diagonal stack; it is usual for these to be built from a rectangular base, which must be of sufficient size. It will be seen that the outer walls and withes of this rectangular base are of 1-brick thickness. The spaces created on the rectangular base where the diagonal stack commences, are filled in with tumbled brick weatherings.

Weathering of chimney stacks passing through sloping roofs

Figure 11.35 shows a horizontal dpc of sheet lead bedded in chimney stack brickwork, and provision of lead flashings between brickwork and roof tiling.

12

External works brickwork

Garden or boundary walling, paving and steps, retaining walls or ramps and planter boxes are all examples of external works brickwork. This use of brickwork is sometimes referred to as 'hard landscaping'.

Bricks and mortar in any of these situations are very much more at risk from frost damage than the cavity walling of a house.

The outer leaf brickwork in Fig. 12.1 is nicely sheltered under a pitched roof. Any wetting from wind-blown rain is on to the outer surface only and can quickly dry out so that bricks are unaffected by frost in winter.

The garden wall shown in Fig. 12.2 is totally exposed to rain on both sides and the top and therefore much more at risk from the effects of frost when bricks are saturated.

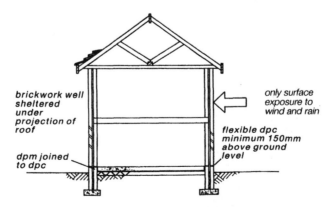

brickwork well
sheltered
under
projection of
roof

only surface
exposure to
wind and rain

flexible dpc
minimum 150mm
above ground
level

dpm joined
to dpc

Figure 12.1 The importance of typical roof projection at eaves in protecting external brickwork

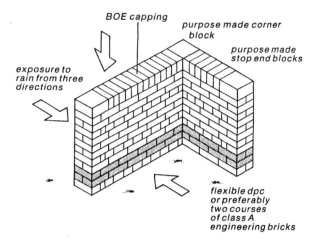

Figure 12.2   Free standing wall with flush brick-on-edge capping

Selection of bricks for external works ____

Bricks that soak up a lot of water are more likely to suffer from the effects of frost when in a saturated state than bricks which absorb very little water. Class 'A' engineering quality bricks, due to their method of manufacture absorb no more water than 4.5% by weight, and so are unaffected when exposed to freezing conditions.

See Chapter 2 for definitions of engineering bricks.

Some facing bricks absorb as much as 20% of their weight in water, and if frozen in this condition, will begin to crumble away at the surface.

Great care must be taken with the choice of brick for external works. Engineering quality or other 'frost proof' bricks must be specified if brickwork is to have long term durability. From an appearance point of view well burned stock bricks blend well with gardens and landscaping. If 'clamp burnt stocks' are specified, only first or second-hard selected bricks should be used for landscape brickwork constructions. Always check with the manufacturer before using any brick for external works.

Mortar _____

The choice of mortar for external works or garden brickwork must be as carefully considered as the bricks, if it also is to be durable.

Group (ii) mortar 1: $\frac{1}{2}$: $4\frac{1}{2}$, will give increased resistance to the effects of frost, if used with hard burnt stocks and Class 'B' engineering bricks.

Group (i) mortar 1: $\frac{1}{4}$: 3 complements Class 'A' engineering bricks, with its higher compressive strength when hardened, and therefore has greater resistance to water penetration (See Chapter 2 for mortar designations).

Brick paving

Paving bricks or 'pavers' are classed as frost proof because they are likely to be damp throughout the year, and are particularly exposed to frost damage in winter.

Method of laying pavers

There are two ways of laying paving bricks:-

(i)   solid bedding on to a previously laid concrete base slab using bricklaying mortar below and between, see Fig. 12.3.
(ii)   flexible bedding. This is the most commonly used and most economical way of laying paving bricks, as it dispenses with the need for a concrete foundation slab and neither is bedding mortar required, see Fig. 12.4.

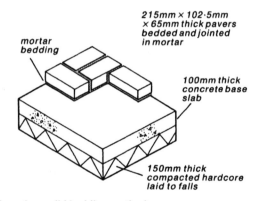

Figure 12.3   Brick paving: solid bedding method

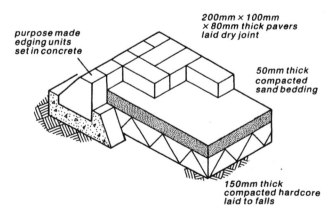

Figure 12.4   Brick paving: flexible bedding method

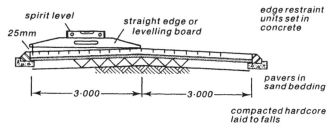

Figure 12.5   Section through paving to show minimum required fall or gradient

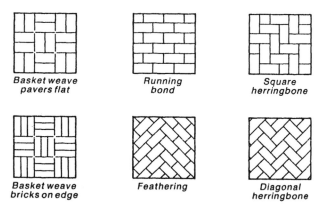

Figure 12.6   Typical patterns for brick paving

When laying bricks by the flexible bed method, pavers are placed upon a carefully levelled layer of sand 50mm thick, and settled in with a mechanical plate compactor. Dry sand is brushed over areas of completed brick paving to fill any gaps. Rainwater will drain down between pavers initially with this method of laying and soak away through the hardcore sub-base.

Pathways, patios and larger areas of brick paving must always be 'laid to falls', whichever bedding method is specified, in order to prevent puddles forming in wet weather. Although initially, rainwater drains away through the 'dry joints' of flexible bedding, these soon become sealed up with dust and dirt and rainwater must 'run-off' paving to falls or gradients which should be not less than 25mm per 3m, see Fig. 12.5.

Various patterns of brick paving are possible as shown in Fig. 12.6, some using 200mm × 100mm size paving units, others using 215mm × 102.5mm purpose made pavers, or frost proof bricks laid 'frog down' or on-edge.

Boundary walls

These are described as 'free standing' because the brickwork derives no support from floors or roof structure as that used in a building.

For this reason Local Authority Building Control Departments' approval is usually required if a wall higher than one metre is to be built next to a public road in order to check that the design is stable.

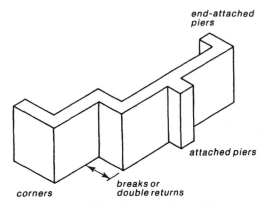

Figure 12.7   Ways of improving stability of free standing walls

Figure 12.8   Serpentine walling

Corners, attached piers and 'breaks' in a free standing wall provide support to resist overturning and failure at the weakest point, where a flexible d.p.c. has been used.

Attached piers must be properly bonded to the wall as shown in Fig. 7.13, page 98.

Damp proof coursing for boundary walls is best provided by using two courses of black or red Class 'A' engineering bricks, bedded and jointed in cement mortar.

Serpentine walling of only 102.5mm thickness stretcher bonded brickwork is a very effective shape to resist overturning, but is only convenient in large parkland situations. See Fig. 12.8, notice also the vertical movement joint, and two courses of Class 'A' engineering bricks as dpc.

The brickwork above and below is solidly bonded to these dpc bricks, as shown in Fig. 12.9. (Note also, two more courses of contrast-colour bricks

Figure 12.9 Two courses of Class A engineering bricks as dpc

mid-height, for decorative effect). The result bonds better than if separated by a flexible dpc, which in a free-standing boundary wall creates a plane of weakness.

Vertical movement joints

The bricklayer must make allowances, when constructing boundary walls, for the expansion and contraction of the brickwork due to changes in air temperature and/or moisture content of the bricks.

A completely straight-joint should be formed throughout the full height and thickness of the brick wall. These vertical movement joints should be 10 to 15mm wide on the face, and are needed at approximately 9-metre intervals to prevent stress from developing in the wall due to expansion. A soft filler of expanded plastic foam strip is built in, to stop mortar bridging the 10 to 15mm space.

To reinforce the obvious weakness caused by these unbonded, vertical joints, special stainless steel 'slip-ties' are built in every fourth course as the work proceeds, as shown in Fig. 12.10, to keep the ends of the walling permanently in line. Standard cross-cavity wall ties must *not* be used at these points.

The expanded plastic foam filler is kept back (or cut back), 12mm from both faces of the brickwork, so that a sealant-mastic can be applied, to seal these movement joints against rain penetration.

It is very important that vertical movement joints are taken right up through any BOE or precast coping or capping on top of the wall. Generally speaking, vertical movement joints should commence at ground level, because the temperature below is practically constant.

Note. Compressed fibre board is unsuitable as a vertical movement joint filler in brick walling. It is too hard, and does not compress easily when the walling expands.

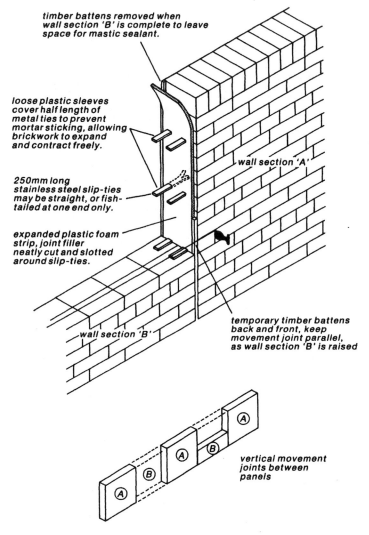

Sequence of construction of wall sections

Figure 12.10   Formation of vertical movement joint

Construction

It is very important that vertical movement joints are truly plumb, straight
and parallel 10 to 15mm width on face. A good way of ensuring these three
important requirements, is to build the wall in alternate sections between
vertical movement joint locations, as shown in Fig. 12.10.

The stopped ends of wall sections (A) can be accurately plumbed and
built to gauge first. When wall sections (B) are infilled, courses being run-in
to the line, using 10 to 15mm thick temporary timber slips to leave truly
straight and parallel vertical joints upon completion of the wall.

Copings and cappings

The top of free standing walls must be protected as effectively as is possible, to prevent downward penetration of rain water. A continuous top covering which looks attractive and acts like a good umbrella will give the best long term protection to a free standing wall and avoid damage to brickwork by frost.

Fig. 12.11 shows the desirable features of a well designed coping. Flush cappings lack some of these important features and so afford less effective protection. See Fig. 12.2.

Continuous lengths of non-ferrous metal or plastic would provide the best sort of protection from rain, due to having very few joints, but these materials are not commonly used.

Lengths of precast concrete, typically 750mm long and saddle back or feather edge in cross section embody the best features of the well designed coping as indicated in Fig. 12.11, but leakage can occur at the mortar joints.

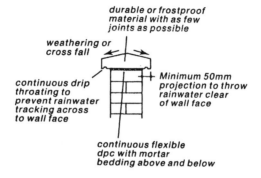

Figure 12.11 Requirements of an effective durable coping

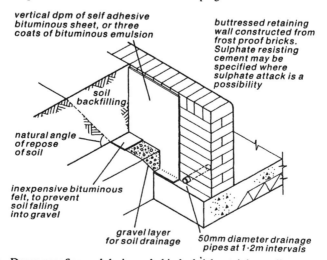

Figure 12.12 Damp proofing and drainage behind a brick retaining wall

Natural stone copings of similar cross section are much more expensive as these are sawn from naturally quarried Portland or other durable stone.

Plain brick-on-edge copings and cappings, although traditional and very commonly used, present many mortar joints through which rain can penetrate, see Fig. 10.19.

All stone, precast concrete and brick copings and cappings should be bedded upon a continuous dpc to arrest any leaking at joints. See Fig. 12.11. Various patented extensions of the basic brick-on-edge are available, which seek to make copings and cappings more secure. See Fig. 12.15.

Brick retaining walls

Any walling intended to hold back soil must be properly designed by a structural engineer, so as to resist overturning from sideways pressure. Even a 215mm thick wall that is only six courses high and intended to hold back a soil bank may begin to topple after a couple of years, see Fig. 12.13.

The line AB in Fig. 12.13 represents a typical natural angle of repose for soil, say 45 degrees. Motorway embankments are frequently seen left at this natural slope with the soil completely stable with grass and bushes growing on them. The angle of repose, of course, depends on the type and condition of the soil.

It is the 'soil wedge' *above* line AB in Fig. 12.13 that presses against the vertical back of a retaining wall, along its full length, which is tending to push over the brickwork. Retaining walls must be able to permanently resist the pressure of this triangular mass of soil and remain plumb.

Any extra load at point C in Fig. 12.13 *i.e.* stacks of materials or vehicles, is called a surcharge, and increases the overturning pressures on the retaining wall.

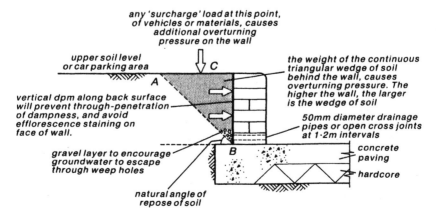

Figure 12.13 Cross section through low brick retaining wall, showing typical sideways pressure from soil behind

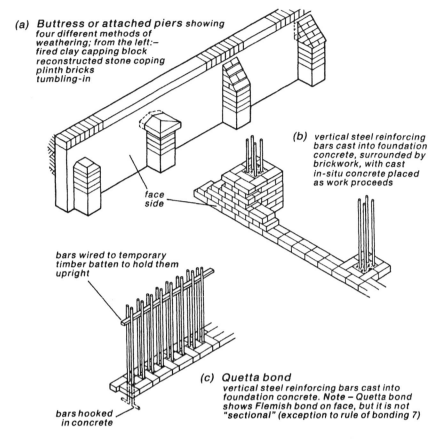

(a) **Buttress or attached piers** showing
four different methods of
weathering; from the left:–
fired clay capping block
reconstructed stone coping
plinth bricks
tumbling-in

face
side

(b) vertical steel reinforcing
bars cast into foundation
concrete, surrounded by
brickwork, with cast
in-situ concrete placed
as work proceeds

bars wired to temporary
timber batten to hold them
upright

(c) **Quetta bond**
vertical steel reinforcing bars cast into
foundation concrete. Note – Quetta bond
shows Flemish bond on face, but it is not
"sectional" (exception to rule of bonding 7)

bars hooked
in concrete

Figure 12.14 Three ways that a structural engineer may consider strengthening a brick
retaining wall

Retaining wall construction

Fig. 12.14. shows a number of ways that a brick retaining wall can be
strengthened against sideways or lateral pressure.

Class 'A' or 'B' engineering quality, or other frost proof bricks should be
specified, using a Group (i) or Group (ii) mortar, to ensure long term
durability.

The soil side of a brick retaining wall should be covered with a dpm
(damp proof membrane), to prevent through-penetration of ground water,
which can cause unsightly efflorescence or other salt staining of the outer
surface. This dpm can be formed from 900mm wide rolls of paper-backed
self adhesive bitumen, pressed firmly on to the brickwork at the soil side of
the wall.

Alternatively, three separate coats of bituminous emulsion paint may be
applied to the back surface of the brickwork to form a vertical dpm, see
Fig. 12.12.

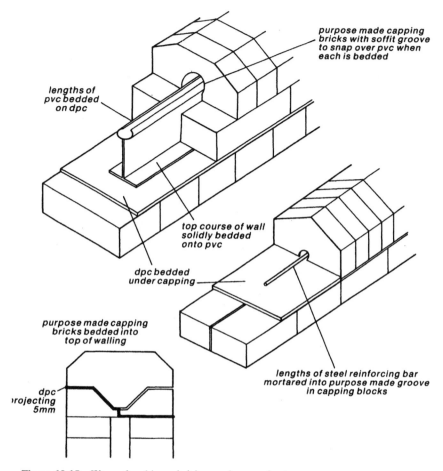

purpose made capping bricks with soffit groove to snap over pvc when each is bedded

lengths of pvc bedded on dpc

top course of wall solidly bedded onto pvc

dpc bedded under capping

purpose made capping bricks bedded into top of walling

dpc projecting 5mm

lengths of steel reinforcing bar mortared into purpose made groove in capping blocks

Figure 12.15   Ways of making a brick-on-edge capping course more secure

Drainage

Small drain pipes should be built into the lowest course of a brick retaining wall, to prevent any build up of static water pressure. A layer of hardcore or gravel behind the retaining wall should be spread between these drainage pipes, before the soil is backfilled, upon completion of the wall, see Fig. 12.12.

13
Special shaped bricks

The bulk of production by brick manufacturers is of plain, basic metric bricks, with a work size of $215 \times 102.5 \times 65$mm. These are produced as solid, frogged or perforated, from clay, sand lime or concrete as described in Chapter 2 of this book.

Manufacturers do, however, make a wide range of other brick shapes, and some examples are shown in Fig. 13.1.

BS 4729:1990 describes these variants as 'Bricks of Special Shapes and Sizes'. The majority are based upon the dimensions of the basic metric brick (see Fig. 13.2), so they will fit in with normal bonding arrangements and vertical gauge. These are given a type number and prefix number in BS 4729, to make ordering a simple matter.

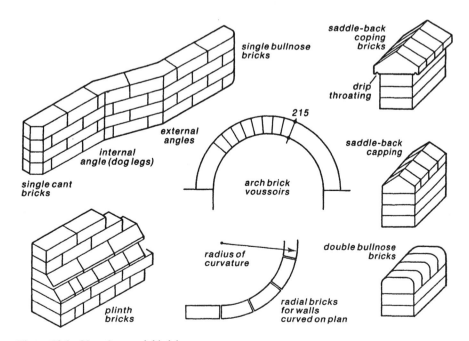

Figure 13.1 Uses for special bricks

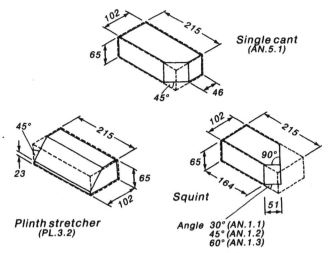

Figure 13.2   Enclosing dimensions of typical special shape bricks

Bricklayers know the most commonly used ones as 'standard specials', intended for the decorative and functional purposes required of brickwork—see Fig. 13.3. They usually refer to all other shapes that might be specified by an architect as 'special specials'. These require detailed drawings to be sent to the brick manufacturer, so that a mould can be made.

Availability of special shaped bricks _____

Although a brick of special shape might be referred to as a 'standard special', it will not necessarily be held in stock by the brick manufacturer or builders' merchant. Since the hundreds of different colours and surface textures of facings each have their own respective range of special shapes, stocking them all would be a very expensive business.

The BS 4729:1990 divides special bricks into ten groups for convenience of classification. Each group has recognisable prefix letters to make the system user-friendly, see Fig. 13.4, which shows a selection from the total range available.

Stop bricks _____

Certain special bricks are shaped so that, for example, a bullnose effect can be changed neatly back to a square corner. Examples of different kinds of 'stop bricks' are shown in use in Figs. 13.5 and 13.6.

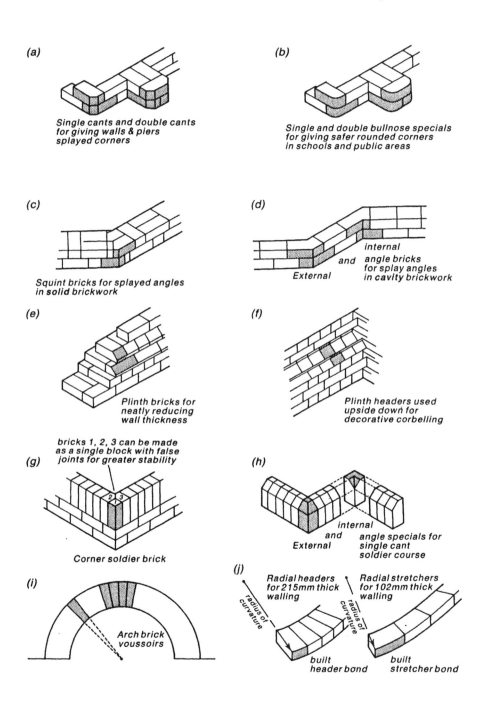

Figure 13.3   Application of special shape bricks

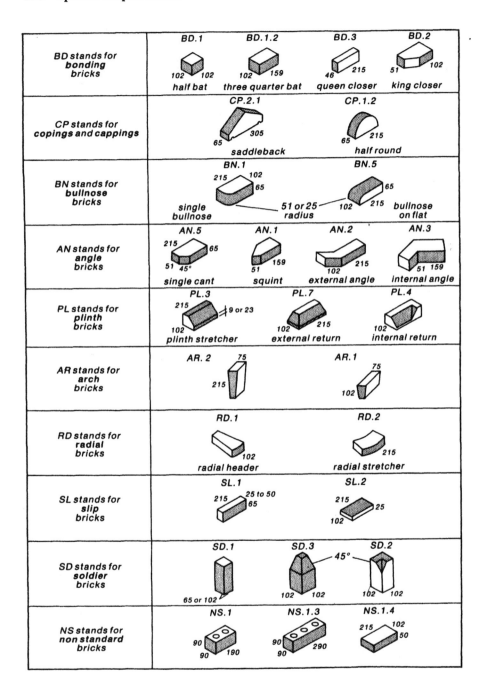

Figure 13.4 BS4729:1990 Classification (Note: dimensions shown in mm are net brick sizes, without joint allowances)

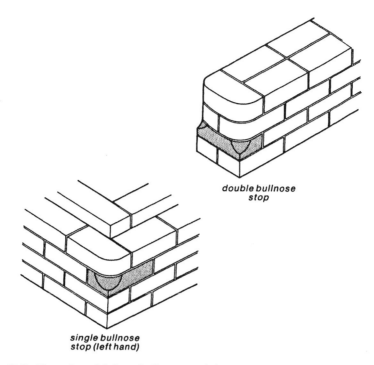

double bullnose
stop

single bullnose
stop (left hand)

Figure 13.5 Uses of special shape bullnose stop bricks

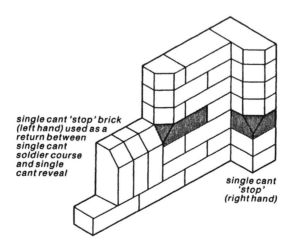

single cant 'stop' brick
(left hand) used as a
return between
single cant
soldier course
and single
cant reveal

single cant
'stop'
(right hand)

Figure 13.6 Uses of special shape cant stop bricks

Bonding with obtuse angle specials————

Squint bricks (AN.1)

Traditionally, squint bricks are produced specifically for bonding obtuse angle quoins in *solid* brick walling (see Fig. 13.7). The dimensions of each face of the squint brick shown, allow ¼ lap to be maintained on face. However, the bricklayer may be called upon to use squint bricks in 102mm thick walling, in some modern building—although this is not good practice because the ½ lap of stretcher bonding is not being maintained around the obtuse angle quoin.

External angle (AN.2) and internal angle (AN.3) specials

These special shaped bricks, illustrated in Fig. 13.3(d), are intended for bonding obtuse angle quoins in stretcher bond (cavity) walling. The face dimensions of each allow ½ lap to be continued around splayed angles.

> *Note*: Internal angle specials (AN.3), sometimes referred to as 'dog legs', are faced on the sides opposite to external angle special bricks (AN.2).

Arch bricks ————————

A trained bricklayer is able to set out an axed arch (see page 145), and cut the required number of voussoirs from basic-size bricks using hammer, bolster and comb hammer.

BS 4729:1990 includes four standard shape arch bricks, however, which manufacturers will mould to a tapering shape, suitable for set spans for semicircular arches of 910, 1360, 1810 and 2710mm, see Fig. 13.8. Arches of any other span or shape will require special shape arch bricks, to obtain parallel mortar joints between voussoirs, and would need to be specially ordered.

Brick on edge quoin blocks, angles and stopped ends ————————

To avoid cutting mitres at the point where a BOE capping turns a corner or obtuse angle, a range of fired clay 'blocks' are made for the purpose. Figure 13.9 shows that these BS 4729 specials are made to suit bullnose and cant as well as plain BOE finish. All are very secure when firmly bedded, providing solid support for intermediate bricks.

'Handing' of special shaped bricks ————

If the single cant special bricks, shown in Fig. 13.3(a) have a smooth face and are solid wirecuts, then they can be used as the 'quoin bricks' in *every* course, simply by being turned over in alternate courses. They do not have a top and a bottom as such.

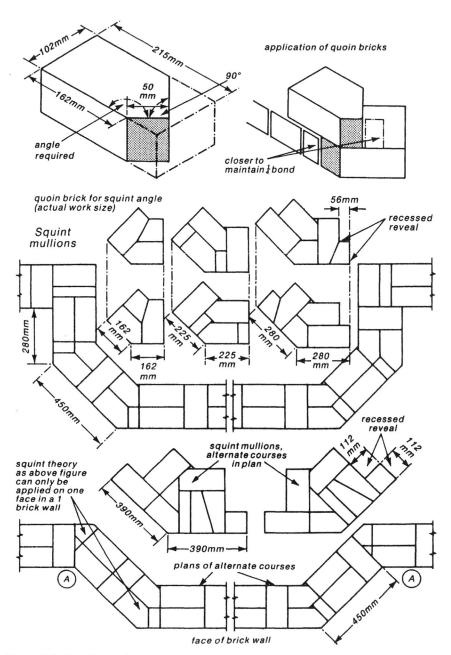

Figure 13.7   Bonding squints

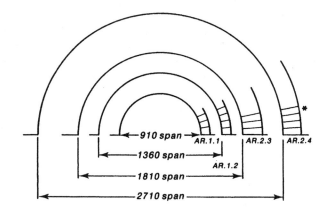

* Extrados dimension with all four types 75mm

Figure 13.8   Arch bricks. Elevation showing four standard spans all available in BS4729 as either tapered headers or tapered stretchers

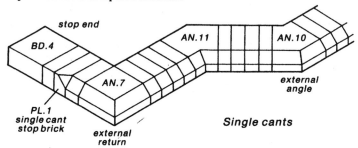

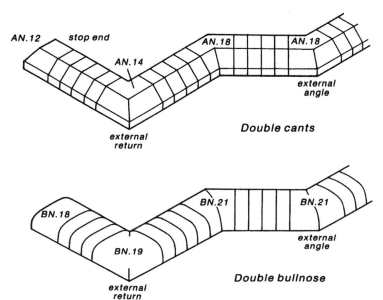

Figure 13.9   BS4729:1990 Brick cappings

Depending upon the method of manufacture, some specials will have a single frog, which should always be laid 'frog-up'. Other facing bricks may be perforated wirecuts with a 'dragwire' surface texture, and these must *not* be turned over either. This is because the surface texture is directional, and the bricks must not be bedded so that rain is retained at the surface.

Therefore, special shape bricks, like those shown in Fig. 13.3(a) to (d) inclusive, have a definite top and bottom, due to the presence of a single frog or a dragwire surface texture. 'Left handed' and 'right handed' versions must be ordered, as indicated in Fig. 13.10.

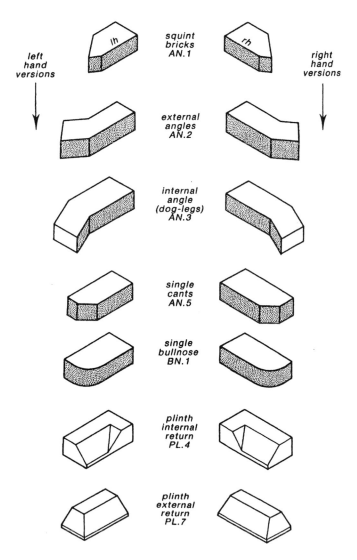

Figure 13.10 Handing of bricks of special shapes and sizes; BS4729:1990 type numbers are included

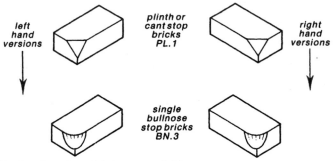

left
hand
versions

plinth or
cant stop
bricks
PL.1

right
hand
versions

single
bullnose
stop bricks
BN.3

Figure 13.11 Handing of special shape stop bricks

Similarly, can you imagine trying to use the single bullnose and single cant stop bricks illustrated in Figs. 13.5 and 13.6 respectively if the change back to a square angle was required on the course *below* that shown in the illustrations? For this reason, stop bricks must also be ordered left and right handed to suit requirements, as indicated in Fig. 13.11.

Radial specials

In a similar way to arch bricks, the BS details some commonly used radial bricks with type numbers that can be quoted when ordering, see Fig. 13.12. If a different radius of curvature is required to construct curved brickwork, then detailed drawings must be sent to the brick manufacturer, from which special moulds can be made.

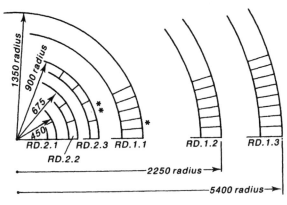

1350 radius

900 radius

675

450

RD.2.1 RD.2.3 RD.1.1

RD.2.2

RD.1.2

RD.1.3

2250 radius

5400 radius

** Extrados dimension 226mm
* Extrados dimension 108mm

Plan views of walling

Figure 13.12 Radial bricks. Plan view showing six standard radius curves, all available in BS4729 as either tapered headers or tapered stretchers

14

Joint finishing

The surface finishing treatment of new facework may be a jointing or pointing operation for bricklayers, and has an important effect upon the finished appearance of brickwork. When looking at stretcher bonded facework 18% of what you see is mortar colour. (See chapter 15 Example 22, page 286).

Jointing is the craft term applied when joints are finished with the same mortar as is being used for the bricklaying, while the work proceeds.

Pointing is the term used to describe the surface finish applied to the cross joints and bed joints of a brick wall when raked out to a depth of approximately 12mm, and filled with a mortar of different colour or texture.

Fig. 14.1 shows a number of different joint finish profiles. 'Weather struck and cut' and 'Tuck pointing' are time consuming and so are always applied as a *pointing* operation some weeks or months after the wall has been built, and the joints raked out in preparation for it. 'Square recessed' finishing can only be done as a *jointing* process. The remaining profiles shown in Fig. 14.1 may be carried out as *either jointing or pointing* operations.

Reasons for joint finishing _____

The purpose of joint finishing is to compact the surface and press the mortar into contact with the arrises of each brick, so as to prevent rain penetration and possible frost damage.

Another reason for jointing or pointing, is to give the exposed mortar surfaces an even appearance overall. The word 'weather' used in any description of jointing or pointing simply means 'sloped', so as to shed or tip rain water off the surface of joints.

Tools _____

Those items in a bricklayer's tool kit used for joint finishing can be a mixture of purchased and home made items as shown in Fig. 14.2.

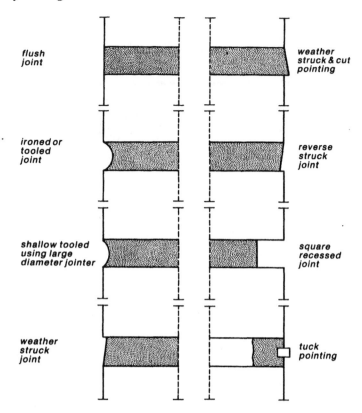

flush joint

weather struck & cut pointing

ironed or tooled joint

reverse struck joint

shallow tooled using large diameter jointer

square recessed joint

weather struck joint

tuck pointing

Figure 14.1 Joint profiles

Mortar mixes

Where facework is to be 'jointed as work proceeds', the bricklaying mortar of course provides the joint finishing colour. 'Ironing in' bricklaying mortar made from a fine grain building sand for example, will leave a smoother surface than where coarser local sand is in use.

If fine grain building sand is used to produce *pointing* mortar, then a 'weather struck and cut' finish will polish up better and may be 'cut' or trimmed more cleanly with the frenchman, than when the sand is coarser.

Cement-rich or 'strong' pointing mortar e.g. BS 3921 Group (i) should be reserved for very dense Class 'A' engineering bricks only. Group (ii) 1: $\frac{1}{4}$: 4 $\frac{1}{2}$ is suited to Class 'B' bricks.

A slightly stronger mix than Group (iii) 1:1:6 is in order for the majority of bricks with compressive strengths between 20 and 40 N/mm^2, e.g. 1:1:5. The reduction in sand improves the 'fattiness' of the mortar, so that it sticks to the pointing trowel. If the fattiness of a pointing mortar needs further improvement, it is better to increase the proportion of lime rather than the cement; see Chapter 2 (e.g. Fig. 2.7).

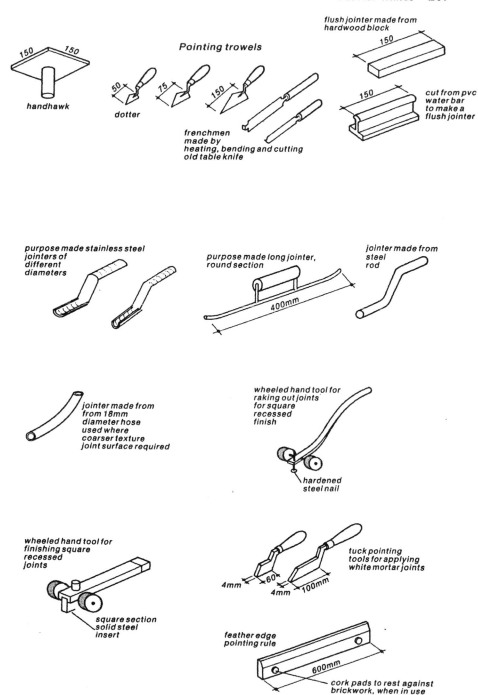

Figure 14.2 Tools for jointing and pointing

Very careful and consistent batching is necessary, with strict control of mix proportions by volume using gauge boxes or buckets each time, if mortar is always to finish up the same colour and strength.

The safest way of producing coloured mortar joints, is to use tinted lime/sand delivered to the site in bulk, to which is added one part cement, each time a batch is required. This way the mortar supplier will have batched lime, sand and pigment at works by weight, which is a more reliable method than site-batching by volume.

Sample panels

Before facework commences, small sample walls of approximately $1m^2$ should be built and jointed or pointed, so that the architect may see the effect of the specified joint finish and its colour. More than one sample panel may be required and each should be allowed to dry out fully before final choices are made.

Choice of joint finish

Flush joints

Flush jointing is usually specified when the architect does not wish to see any shadow lines cast by the joint finish. Another reason for its use is where a 'matt' joint surface is preferred, instead of one smoothed or struck by a steel trowel or metal jointer.

Although this is a very simple looking joint finish, it is not as easy as it looks to get the mortar surface truly flush—not slightly 'dished'. The hardwood block or PVC slip shown in Fig. 14.2 must be used to smooth and compact the surface mortar, but must not leave it pressed further back than the face of the bricks; see 'Technique—jointing' below.

Flush jointing with handmade or stock bricks can make joints appear wider than they really are, due to the rounded arrises of these bricks.

Weather struck and cut pointing

This can make courses of irregular shaped bricks appear straighter than they are; see 'Technique—pointing' below.

Possibly the most commonly specified joint finish where new brickwork has been raked out, and is to be pointed at a later date or when old brickwork needs repointing. This joint finish ensures that rain is tipped off the exposed bed joint surfaces and arris of each brick; but it is not practical to use as a 'jointing' operation as work proceeds. The *additional* mortar which must be applied to permit neat 'cutting' of cross and bed joints (see Figs. 14.4 and 14.5), would delay economical bricklaying progress.

Weather struck joints

This 'weathered' joint finish *can* be carried out as a jointing operation while work proceeds, for no delaying 'cutting' is required. It includes the advantages of a sloping surface to the exposed bed joints, to improve weather resistance.

The depth of indenting the left-hand side of cross joints and the top of bed joints of this profile should not be deeper than the trowel blade thickness. Care must be taken to see that both left handed and right handed bricklayers indent cross joints on the LHS only.

Bricklayers develop an individual style when performing this jointing operation; some 'strike' the mortar joints as they lay each brick, others complete a whole course and then proceed to strike-up the joints. In either case, it is advisable to use the *brick trowel* for struck jointing, rather than a pointing trowel.

Reverse struck joints

This joint finish is used for internal wall surfaces of common brickwork that are to be coated with emulsion paint. Joints are struck the opposite way to weather jointing, and not trimmed or 'cut'. Walls have a flat even appearance when viewed from below, free of joint shadows. This trowelled finish performed with the bricklaying trowel rather than a pointing trowel, should be as flat as possible to avoid ledges which act as dust traps.

Ironed or tooled joints

This is the most common joint finish, allowing the bricklayer to disguise slight chips or imperfections on the arrises of bricks so as to avoid unnecessary waste.

Care should be exercised to see that all the bricklayers in a gang use the same diameter jointers. Smaller diameters give a deeper joint profile. Larger diameter jointers give a shallower profile.

Irregular depth ironing of joints will cast different shadow effects, and cause variations in the appearance of completed elevations of facework. (See 'Technique—jointing' below).

Square recessed jointing

This finish gives a strong shadow effect, but exposes every chip or imperfection on the arrises of bricks. As an external joint finish, must only be used with frost resistant bricks, because bed joints are not weathered, rain can lodge on recessed arrises giving the risk of frost damage.

The depth gauge on each bricklayers wheeled jointer must be set the same e.g. 6mm or 8mm when raking out mortar joints. All, square-recessed joints must then be polished directly with the solid steel insert of the wheeled jointer to leave the recessed mortar surfaces smooth, square and clean, see Fig. 14.2.

Tuck pointing

A now practically obsolete and extremely time consuming system of pointing which applies false cross joints and false bed joints only 3mm wide to the face of previously flush jointed brickwork. It developed in eighteenth century Britain as a next best thing, if you could not afford genuine gauged brickwork built with red rubbers and fine white lime putty joints. In modern practice it is generally only used where there is a requirement to match existing work during refurbishment of period buildings.

The false white mortar joints, referred to as 'the strip' by the few modern tuck pointers, is superimposed on a flush joint, called 'the stopping'.

Stopping mortar for the flush joint is either coloured to suit the bricks in the wall, or in the case of old buildings, ordinary mortar is used, and the whole wall face is colour washed before 'the strip' is applied.

Tuck pointing mortar consists of a mix of lime putty and silver sand 1:3, although modern tuck pointers frequently include $\frac{1}{2}$ part of white cement to make the strip more durable. The flush pointed stopping mortar should be keyed or grooved on the previous day so that the white tuck pointing is firmly attached, (see Fig. 14.1)

Gauge rods are necessary to ensure exact spacing of these joint lines marked in the flush stopping mortar, and the bond pattern accentuated with perfectly plumb perpends.

The white mortar is loaded carefully along one edge of the pointing rule, from which it is pressed on to the face of the flush joint stopping, using purpose-made tuck pointing tools, (see Fig. 14.2)

Bed joints must be applied first, (the opposite way round compared to weather struck and cut pointing), and trimmed top and bottom using a feather edge pointing rule and frenchman. Cross joints are applied last and similarly trimmed with pointing rule and frenchman.

Technique—jointing

Whichever finish is to be used, timing is very important when jointing-up as the bricks are laid throughout the working day. The mortar between bricks should be allowed to stiffen up just enough due to brick suction, so that the jointing tool can pass smoothly and cleanly. Too soon and the mortar smears and does not leave a smooth profile. Left too long before jointing, and heavy pressure on the jointer leaves black metal marks on the dried mortar face.

With the single exception of tuck pointing, whether jointing or pointing, always do the cross joints first, followed by bed joints, each time you stop bricklaying to joint-up.

Brushing-off with a soft bristle hand brush, to remove any loose crumbs of mortar, should be left until the end of the day. Brush lightly and on no account leave bristle marks in the mortar face. Better to leave brushing until the following morning than to risk marking the joints. Take particular care when jointing face brickwork at those points shown in Fig. 14.3.

Technique—pointing

This craft operation, carried out some weeks or months after the wall has been built, requires patience and is a skill which takes time to develop. The

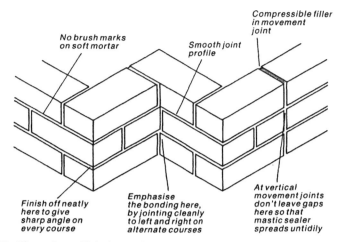

No brush marks
on soft mortar

Smooth joint
profile

Compressible filler
in movement
joint

Finish off neatly
here to give
sharp angle on
every course

Emphasise
the bonding here,
by jointing cleanly
to left and right on
alternate courses

At vertical
movement joints
don't leave gaps
here so that
mastic sealer
spreads untidily

Figure 14.3 Fine points of jointing and pointing

joint finish commonly specified for *pointing* brickwork is weather struck and cut, (see Fig. 14.1)

1 Always start at the very top of the walling to be pointed.
2 Remove any obvious hardened crumbs of mortar clinging to the wall face when the joints were raked out some weeks or months ago.
3 Brush the whole lift of brickwork using a stiff bristle hand brush.
4 Wet the wall face generously if the bricks are very absorbent, less generously if the bricks have a lower suction rate. (See item 14 below for walls with alternate bands of Class 'A' and absorbent facings).
5 Load the hand hawk with mortar flattened out to approximately 10mm thickness.
6 Using the small pointing trowel or 'dotter', pick up joint sized pieces of mortar from the hawk and press carefully and firmly into cross joints, but see also item 14 below.
7 Completely fill each cross joint with a second application if necessary, polish the mortar surface and indent on the LHS
8 After completing approx. $\frac{1}{2}$m^2 of cross joints, cut or trim the RHS of all these joints in the manner shown in Fig. 14.4, so that all look the same width on face.

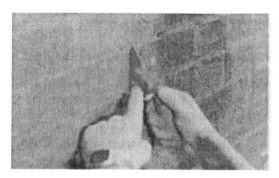

Figure 14.4 Weather struck and cut pointing: cutting the right hand side of cross joints

9 Using a longer pointing trowel, pick up joint sized pieces of mortar from the hawk and commence pointing one bed joint, pulling the loaded trowel up to the last piece of mortar applied each time.

10 After filling a 500mm length of bed joint, polish it with the pointing trowel and indent the top.

11 When half a dozen bed joints have been pointed in this way, cut or trim the bottom edge of each one as shown in Fig. 14.5 using frenchman or tip of pointing trowel together with feather edge pointing rule.

The amount of mortar to be left projecting from the wall face after trimming or cutting joints should not exceed thickness of the trowel blade.

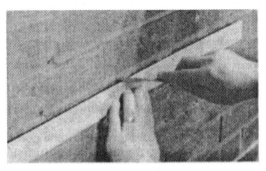

Figure 14.5 Weather struck and cut pointing: trimming the bottom of bed joints with straight-edge and frenchman

12 Sensibly adjust areas of pointing to suit drying conditions of bricks and weather, so that joints cut cleanly. Too early, and the mortar will not fall away cleanly when trimmed. Too late, and joint edges will crumble and not leave a clean straight line when cut.

13 Brush very lightly at the end of the day or preferably on the following day if there is the slightest risk of marking the sharp cut edges of the pointing.

14 If a wall has alternate bands of Class 'A' and absorbent facing bricks, after wetting the whole wall, point absorbent facings first. When the wall has dried off, return and point the Class 'A' bands.

Repointing old brickwork

Chimney stacks and parapet walls are usually the first parts of a brick building which will need repointing after twenty or thirty years exposed to wind, rain and frost. Raking out and repointing can give such brickwork a new lease of life, providing that it is otherwise structurally sound.

1 Rake out all old mortar, using bolster and comb chisel, to a depth of 15–18mm, taking care not to damage brick arrises unnecessarily.

2 Remove all old mortar and dust by vigorously brushing brickwork with a stiff bristle brush.

3 Thoroughly wet the brickwork, and when surface is dry, commence repointing from the top down.

4 Continue as outlined in this Chapter, sub-heading 'Technique— pointing'.

Summary _____

The great merit of surface finishing mortar as a 'jointing' process is that the joint profile is an integral part of the mortar bed and there is no possibility of failure through insufficient adhesion between the main mortar bed and the surface finish.

Failure of pointing, i.e. its separation from the main mortar bed and consequent falling away, is caused by careless raking out and lack of suitable preparation. To overcome possible failure, joints in new brickwork should be raked out to a depth of at least 12mm as shown in Fig. 14.6 and not as shown in Fig. 14.7.

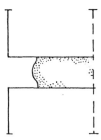

Figure 14.6   Correct way of raking out joints in preparation for repointing

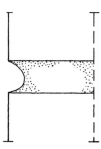

Figure 14.7   Incorrect, careless way of raking out, that will lead to pointing failure due to poor adhesion

15
Calculations

A bricklayer should know how to work out the number of bricks needed to build a wall. This is a very simple matter using a pocket calculator. Do not panic at the very mention of arithmetic. Let the calculator take the strain and deal with the decimal point!

Method A

There is nothing wrong with counting up the number of bricks needed for one course—then multiplying by the total number of courses in the wall to get the answer.

   This method works well for stretcher bond walling, but becomes difficult when applied to more complicated shapes that have doors and windows and also with cavity walling.

Method B

This method is based on finding the surface area of the wall. It makes deductions for doors and windows easy. Use the calculator to add, subtract and multiply. Do not strain the brain!

UNITS of MEASUREMENT—Length in metres (m), sometimes called *lineal* metres
Area in m^2, (square metres)
Volume in m^3, (cubic metres)

Example 1

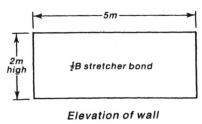

5m

2m high

½B stretcher bond

Elevation of wall

Surface area = $5 \times 2 = 10\text{m}^2$ (square metres)

You need 60 bricks to build one m^2 (1m^2) of ½B thick (102mm) stretcher bond walling.

If the surface area of the wall is 10m^2, then you need $10 \times 60 = \underline{600 \text{ bricks}}$.

Example 2

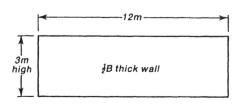

Elevation of stretcher bond

How many bricks needed for this wall?

60 bricks per m^2
Two lines of working method:

Surface area $= 12 \times 3 = 36$m^2

Number of bricks needed $= 36 \times 60 = \underline{2160 \text{ bricks}}$

STOP
When doing any calculation always write down your working method as
shown by the two lines in Example 2. This is worth it, because you can then
check what you did. It will also ensure that you do not get in a muddle with
the decimal point.

Walls 1B thick (215mm) need 120 bricks to build one m^2 (1m^2), and the
same working method can be used as shown in Example 2.

Example 3

How many bricks needed for this wall?

120 bricks per m^2
Two lines of working method:

Surface area $= 16 \times 4 = 64$m^2

Number of bricks needed $=$
   $64 \times 120 = \underline{7680 \text{ bricks}}$

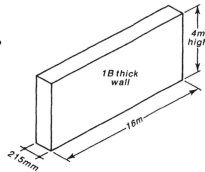

Walls are not always straight elevations, and buildings have corners, but
the same method of calculation can be used.

Example 4

How many bricks are required to build
the wall shown?

60 bricks per m^2
Three lines of working method here:

Total length of wall $= 7 + 5 = 12$m
Surface area $= 12 \times 3 = 36$m^2

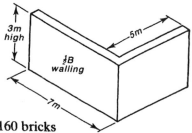

Number of bricks required $= 36 \times 60 = \underline{2160 \text{ bricks}}$

STOP

If calculations are not a happy thought for you, it is always a good idea to retrace your steps. Go back to Example 1. Just change the dimensions of that wall, invent your own figures, and then work out how many bricks will be needed this time.

Repeat what you have just done, but with Examples 2, 3, and 4, to calculate the number of bricks required for each example, but with different wall dimensions.

Example 5

How many bricks needed to build this wall?

120 bricks per m^2
Three lines of working method:

Total length = 8.5+4.5 = 13m
Surface area = 13 × 5 = 65m^2

Number of bricks needed = 65 × 120 = 7800 bricks

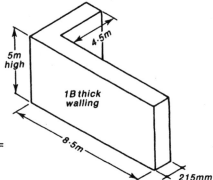

Example 6

How many bricks required to build this wall?

120 bricks per m^2
Three lines of working method:

Total length = A+B+C = 8.7+4.8+9.5 = 23m
Surface area = 23 × 4.5 = 103.5m^2

Number of bricks required = 103.5 × 120
= 12 420 bricks

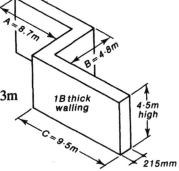

STOP

For additional practice, repeat Examples 5 and 6 after changing the dimensions, invent your own, and assume that the walls are only 102mm thick.

Blockwork

The same method can be used to find out how many *blocks* are needed to build a wall. There are 10 standard size blocks (440 × 215mm on face), needed for one m^2 of walling.

Example 7

Calculate how many standard size blocks are required to build this wall.

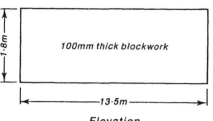

100mm thick blockwork

├─────────13·5m─────────┤

Elevation

10 blocks per m^2
Two lines of working method:

Surface area $= 13.5 \times 1.8 = 24.3\text{m}^2$

Number of blocks $= 24.3 \times 10 = \underline{243 \text{ blocks}}$

Example 8

How many standard size blocks are needed to build this wall?

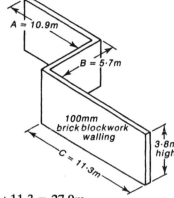

A = 10.9m

B = 5·7m

100mm brick blockwork walling

3·8m high

C = 11·3m

10 blocks per m^2
Three lines of working method:

Total length of wall $= A+B+C = 10.9+5.7+11.3 = 27.9\text{m}$
Surface area $= 27.9 \times 3.8 = 106.02\text{m}^2$

Number of blocks $= 106.02 \times 10 = 1060.2$

Rounded off to $\underline{1061 \text{ blocks}}$

STOP
When ordering blocks for the inner leaf of cavity walls or for partition walls, take care to make clear the thickness you want, 100, 140, 150, 190 or 215mm, as well as type, compressive strength, density and surface finish.

Mortar

Allow .02m^3 (cubic metre) of ready-mixed mortar to build one m^2 (1m^2) of stretcher bonded brickwork. This is an average figure, bearing in mind that some bricks have frogs and others are perforated.

Table 15.1 (p. 291) shows approximately how much ready mixed mortar is required for different wall thicknesses in brick and block.

Example 9

Calculate how much mortar will
be needed to build the brick
wall shown.

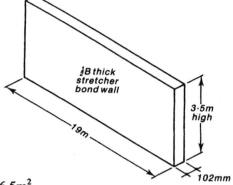

½B thick
stretcher
bond wall

3·5m
high

19m

102mm

.02m^3 of mortar per m^2 of walling.
Two lines of working method:

Surface area of wall = 19 × 3.5 = 66.5m^2

Volume of mortar required = 66.5 × .02 = 1.33m^3

STOP

For additional practice, go back to calculate how much mortar would be
needed to build the wall in Example 4.

Example 10

Calculate how many bricks and how
much mortar will be required to
build this wall junction.

120 bricks and .05m^3 of mortar per m^2

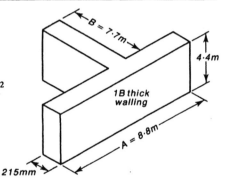

B = 7·7m

4·4m

1B thick
walling

A = 8·8m

215mm

Four lines of working method:

Total length = A+B = 8.8+7.7 = 16.5m
Surface area = 16.5 × 4.4 = 72.6m^2
Number of bricks = 72.6 × 120 = 8712 bricks

Volume of mortar = 72.6 × .05 = 3.63m^3

Example 11

How many standard
size *blocks* and how much
mortar will be needed
to build the block wall
shown?

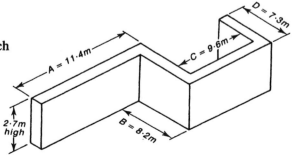

D = 7·3m

A = 11·4m

C = 9·6m

2·7m
high

B = 8·2m

10 blocks and .01m^3 mortar per m^2

Four lines of working method:

Total length = A + B + C + D
$$= 11.4 + 8.2 + 9.6 + 7.3 = 36.5\text{m}$$
Surface area = $36.5 \times 2.7 = 98.55\text{m}^2$

Number of blocks = $98.55 \times 10 = 985.5$
Rounded off to <u>986 blocks</u>

Volume of mortar = $98.55 \times .01\text{m}^3 = .9855\text{m}^3$
Rounded off to <u>1m^3</u>

Rounding off

With building calculations, it is customary to 'round off' decimal parts, so as to tidy up quantities of materials into whole numbers where possible. For example, it is not practical to order half a bag of cement, nor worry about 0.2 of a cubic metre of a bulk item like trench excavation.

It is sensible to 'round up' with building materials, as site handling can result in damage or loss even in excess of the percentage allowances for cutting and waste. Taking Example 11, 1m^3 (1 cubic metre) is a less cumbersome figure than 0.9855m^3 and involves fewer numbers with which to make mistakes. Such rounding-off would not be wise if you were dealing with gold dust or diamonds!

Openings

Most walls and buildings have door and window openings. This means that you must reduce your order for bricks and blocks, depending upon the size and the number of openings.

The best way is to deduct the surface area of an opening from the overall wall area, before the final step of calculating the number of bricks or blocks needed. (See Example 12).

Example 12

Calculate how many bricks will be required to build this wall, making the necessary deduction for the opening.

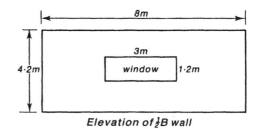

Elevation of ½B wall

60 bricks per m^2
Four lines of working method:

Overall surface area of wall = 8×4.2 = 33.6m^2
Area of window opening = 3×1.2 = 3.6m^2
Nett area of walling = 30m^2

Number of bricks needed = $30 \times 60 = $ <u>1800 bricks</u>

Example 13

How many bricks needed to build the wall shown, making allowances for openings?

60 bricks per m^2

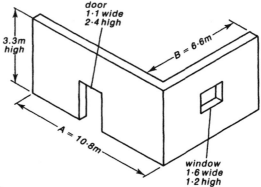

Five lines of working method:

Total length of wall   = A + B   = 10.8 + 6.6 = 17.4m
Surface area   = 17.4 × 3.3   = 57.42m^2

Area of door opening   = 1.1 × 2.4   = 2.64m^2
Area of window opening = 1.6 × 1.2   = 1.92m^2
                Total   = 4.56m^2 4.56m^2

Nett area of walling   = 52.86m^2

Number of bricks required = 52.86 × 60 = 3171.6
                Rounded off to 3172 bricks.

STOP
For additional practice, calculate how much ready mixed mortar will be required to build each wall shown in Examples 12 and 13. (Follow lines of working method given in Example 9).

Cavity walling _____

When working out quantities of materials for cavity work, each leaf of the walling must be dealt with on its own, because you need separate totals for ordering bricks and blocks. (See Example 14).

Example 14

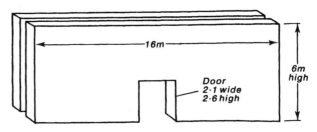

Elevation of 265mm thick cavity wall

Calculate how many facing bricks and blocks will be required to build the cavity wall shown, making due allowances for the opening.

60 bricks and 10 blocks per m^2
Five lines of working method.

Overall surface area	$= 16 \times 6$	$= 96m^2$
Area of opening	$= 2.1 \times 2.6$	$= 5.46m^2$
Nett area of walling		$= 90.54m^2$

Number of facings $= 90.54 \times 60 = 5432.4$
Rounded off to <u>5433 bricks.</u>

Number of blocks $= 90.54 \times 10 = 905.4$
Rounded off to <u>906 blocks.</u>

Method C. The centre line method _____

When walls have many more returns than Example 11, it can be difficult to ensure that corners are not measured twice over. Professional Quantity Surveyors make use of the 'Centre Line Method' to get over this problem when measuring up walls.

This simple but effective idea means that you first work out the total length on plan, of the centre line, of all the intersecting walls of a building having the same thickness. This figure is then used to calculate the total area of walling in one single sum. (See Example 15).

Example 15

Total length *of dotted centre line*

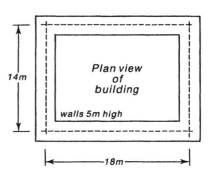

$= (2 \times 18)$ $+ (2 \times 14)$
$= 36$ $+ 28$
$= \underline{64 \ metres}$

Total area of all four walls $= 64 \times 5 = \underline{320m^2}$

The arrowed dimensions shown on all of the foregoing Examples have been indicated so that 'double measurement' of corners has not taken place.

The centre line method of measurement for solid walling, as illustrated in Example 15 can be easily applied to cavity walling.

The first step is to find the total length on plan of the centre line of each leaf of brick masonry.

The longer centre line of the outer leaf will be used to calculate the number of facings needed. The shorter centre line of the inner leaf used to calculate the number of blocks required.

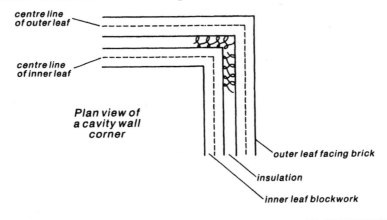

centre line
of outer leaf

centre line
of inner leaf

Plan view of
a cavity wall
corner

outer leaf facing brick

insulation

inner leaf blockwork

STOP. Perimeter.

The perimeter of any shape is simply the total distance around it. A simple building shape, rectangular on plan like Example 16 has an external perimeter and an internal perimeter.

Example 16

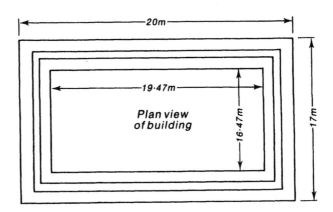

20m

19·47m

Plan view
of building

16·47m

17m

External perimeter = (2 × 20m) + (2 × 17m)
                   = 40m   + 34m   = 74 metres.

Internal perimeter = (2 × 19.47) + (2 × 16.47)
                   = 38.94   + 32.94   = 71.88 metres.

Using the perimeter makes working out centre line lengths easy. (The perimeter is also useful when calculating total surface area for repointing or rendering a building.) Remember, the perimeter is just a plain measurement of length, lineal metres.

Example 17

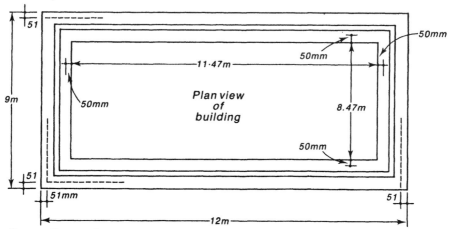

Centre line perimeter of *outer leaf*
= 12m − .102m = 11.898m × 2 sides = 23.796m
  9m − .102m =   8.898m × 2 sides = 17.796m
            Total = 41.592m

Centre line perimeter of *inner leaf*
= 11.47m + .100m = 11.570m × 2 sides = 23.14m
  8.47m + .100m =   8.570m × 2 sides = 17.14m
            Total = 40.28m

Example 18

Assume that the walls of the building shown in Example 17 are 4.5m high. Calculate the number of facings required to build the outer leaf, and number of standard blocks needed for the inner leaf, making use of centre line perimeters. (Forget any door or window openings for this Example).

60 bricks per m²
Five lines of working method:

Total area of *outer* leaf   = Centre line × 4.5m
                                = 41.592m × 4.5m
                                = 187.164m²

Number of facings   = 187.164 × 60   = 11 229.84
Rounded off to 11 230 bricks.

10 blocks per m²
Five lines of working method.

Total area of *inner* leaf   = Centre line × 4.5m
                                = 40.280m × 4.5m
                                = 181.26m²

Number of blocks   = 181.26 × 10 = 1812.6
Rounded off to 1813 blocks.

Example 19

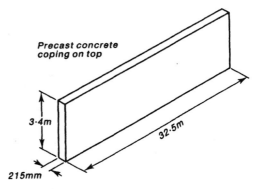

Precast concrete
coping on top

3·4m

32·5m

215mm

Calculate the total area for repointing this brick wall, back, front and ends.

Perimeter
$$= (32.5 \times 2) \quad +(.215 \times 2)$$
$$= 65m \qquad\qquad +.43m$$
$$= 65.43 \text{metres}$$

Total area of repointing $= 65.43 \times 3.4$
$$= 222.462 \text{m}^2$$

Rounded off to $\underline{223\text{m}^2}$

Example 20

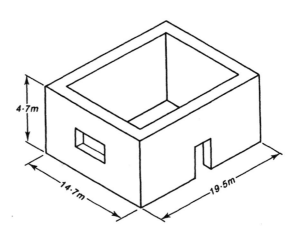

4·7m

14·7m

19·5m

Calculate the total area of raking out joints and repointing external walls of the building shown. (Ignore door and window openings for this example.)

External perimeter
$$= (19.5 \times 2) \qquad +(14.7 \times 2)$$
$$= 39m \qquad\qquad +29.4m$$
$$= 68.4 \text{ metres}$$

Total area of repointing $=$ External perimeter $\times 4.7m$
$$= 68.4 \times 4.7$$
$$= 321.48 \text{m}^2$$

Rounded off to $\underline{322\text{m}^2}$

STOP

For extra practice repeat calculations Examples 19 and 20, after changing the dimensions, invent your own.

Division of facings and commons _____

The outer, 102mm thick leaf of cavity work and other ½B walling is usually built stretcher bond, so that all bricks required will be facings.

With 215mm and thicker solid brickwork faced on one side only, some bricks will need to be less expensive commons to back up the facing bricks.

With 215mm thick Flemish bond walling, ⅔ will be facings and ⅓ commons. With 215mm thick English bond walling, ¾ will be facings and ¼ commons.

If 215mm thick walls are built to Header bond, then of course all bricks ordered should be facings. This information is summarised in Table 15.1.

Example 21

Assume that the wall shown in Example 7 is to be 215mm thick Flemish bonded brickwork instead of blockwork. Calculate the number of facings and commons that will be needed, in separate totals.

80 facings per m² ⎫ From Table 15.1
40 commons per m² ⎭ columns 1 and 2

Three lines of working method:

Surface area $= 13.5m \times 1.8m = 24.3m^2$

Number of *facings* $= 24.3 \times 80 = \underline{1944}$
Number of *commons* $= 24.3 \times 40 = \underline{\;972}$

Percentage for cutting and waste _____

All of the foregoing examples of calculating requirements for bricks, blocks and mortar have resulted in exact or nett quantities. Due to breakages and waste when cutting and handling bricks and blocks and when using mortar, it is necessary to increase nett orders by 5%.

Therefore as an additional stage at the end of each calculation, 5% extra should be added on, using the % button on your calculator, for the purposes of ordering.

Referring to Example 3

Nett amount of bricks required $= 7680$
Add on 5% for cutting and waste $= \;\;384$
TOTAL $= \underline{8064}$ bricks to be ordered.

STOP

Go back to Example 18 and calculate the number of bricks and blocks that should be ordered after allowing 5% for cutting and waste.

Example 22

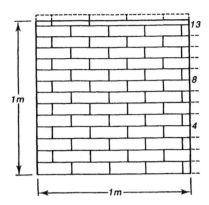

To calculate the area of mortar joints (say to assess the effect of mortar colour on finished walling) showing on the surface of one square metre of stretcher bonded brickwork.

For the purposes of this calculation, assume that the elevation above is 'stack bonded' then:

Bed joint mortar showing $= 1m \times .01 \times 13 = 0.13m^2$
Cross joint mortar showing $= 1m \times .01 \times 5 = \underline{0.05m^2}$
$$\text{Total} = \underline{0.18m^2}$$

Expressed as a % of $1m^2$ $= \dfrac{0.18m}{1m^2} \times 100$

$$= \underline{18\% \text{ of surface viewed}}$$

is mortar colour. The remaining 82% only, is brick colour.

Additional applications of the centre line method

In addition to bricks, blocks and mortar, a bricklayer may be asked to work out quantities of cavity insulation, dpc, foundation concrete or even excavation to trenches.

Each such calculation is very simple, if you find out the total length of the centre line of the walling first.

Example 23

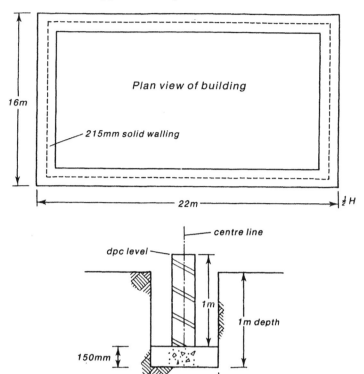

Plan view of building

16m

215mm solid walling

22m

½ H

centre line

dpc level

1m

1m depth

150mm

700mm spread

Centre line perimeter = 22m − .215m = 21.785m × 2 sides = 43.57m
16m − .215m = 15.785m × 2 sides = 31.57m
Total = 75.14m

The dotted Centre line perimeter shown on the drawing is exactly 75.14m long. This dotted line is the exact centre line of the trench excavation, concrete strip foundation, sub-structure brickwork and dpc, all with a nett length of exactly 75.14m, that can be used in four separate calculations.

Trench excavation

Volume = Centre line × width × depth
= 75.14   × .700   × 1m

= 52.598m^3
Rounded off: 53m^3

Strip foundation concrete

Volume   = Centre line × spread × depth
= 75.14   × .700   × .150m

= 7.890m^3
Rounded off: 8m^3

Sub-structure brickwork

Area
$$= \text{Centre line} \times \text{height}$$
$$= 75.14 \qquad \times 1\text{m} \qquad = 75.14\text{m}^2$$

Number of bricks
$$= 75.14 \times 120 = 901.68$$

Rounded off 902 bricks

Dpc

75.14m plus extra for laps.

STOP

Go back to Example 15. Calculate the volume of excavation to form one metre deep by 750mm wide foundation trenches for that Example.

Rounding-off: a warning _____

Rounding up numbers that represent quantities of building materials, when carrying out building calculations, should only be done with the *final* figure, or last result.

Do not round up two or three times during the working stages of a calculation, otherwise the final quantity could be excessively large, and become what is called an 'accumulated error'.

Circles and triangles _____

In addition to those shapes of walls and buildings used as examples in this chapter so far, there are two others that the bricklayer will encounter in construction, which require measurement for setting out and calculation; the triangle and the circle, or parts of it.

Equal gable

Mono pitch roof

North light roof

Examples of triangular shape gable end walls

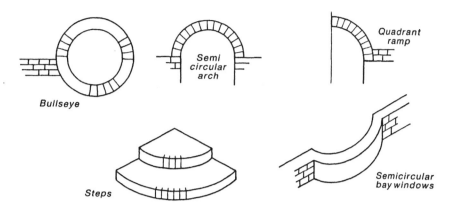

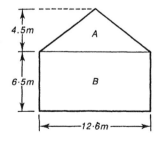

Examples of circular shapes or parts of a circle used in construction

Example 24

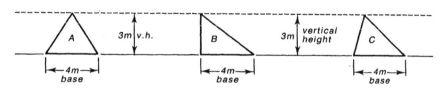

The surface area of any triangle can be calculated as follows, whether shaped like A, B or C:

Surface area = (base × vertical height) ÷ 2
= (4m × 3m) ÷ 2 = 6m²

The surface area of all three triangles is 6m², despite their different shapes.

Example 25

Calculate the surface area of the gable end wall, for repointing.

Total surface area = A + B
Area of A   = (base × v.h) ÷ 2
           = (12.6 × 4.5) ÷ 2
           = 56.7 ÷ 2   = 28.35m²
Area of B   = 12.6 × 6.5   = 81.9m²
Total surface area   = 110.25m²

The area of any circular shape can be calculated from the simple formula πR^2 (pronounced pi, R, squared).

Some pocket Calculators have a button marked π, if not, then use figures 3.14 (the value for pi never changes).

The letter 'R' in the formula stands for radius of the circle.

Example 26

Calculate the surface area of the shaded part of this bullseye.

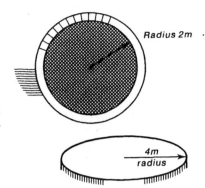
Radius 2m

Surface area $= \pi R^2$
$= \pi \times R \times R$
$= 3.14 \times 2m \times 2m = \underline{12.56m^2}$

Example 27

Calculate the surface area of this circular paved space.

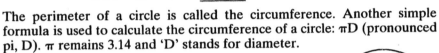
4m radius

Surface area $= \pi R^2$
$= \pi \times R \times R$
$= 3.14 \times 4m \times 4m = \underline{50.24m^2}$

The perimeter of a circle is called the circumference. Another simple formula is used to calculate the circumference of a circle: πD (pronounced pi, D). π remains 3.14 and 'D' stands for diameter.

Example 28

Calculate the circumference of the circle shown.

5m diameter

Circumference $= \pi D$
$= \pi \times \text{Diameter}$
$= 3.14 \times 5m \quad = \underline{15.70 \text{ metres}}$

STOP
Remember that the circumference of a circle is just plain measurement of length, in lineal metres. Surface area is measured in m^2.

Table 15.1 Materials per square metre of walling

	Stretcher bond			English bond			Flemish bond			English garden wall bond			Flemish garden wall bond			Header bond		
	F	C	$M(m^3)$	F	C	$M(m^3)$	F	C	$M(m^3)$	F	C	$M(m^3)$	F	C	$M(m^3)$	F	C	$M(m^3)$
½B 102mm thick	60		.02															
1B 215mm thick	60*	60*	.05	90	30	.05	80	40	.05	73†	47	.05	67†	53	.05	120		.047
1½B 327mm thick				90	90	.08	80	100	.08	73	107	.08	67	113	.08			
2B 440mm thick				90	150	.11	80	160	.11	73	167	.11	67	173	.11			
100mm standard size blocks		10	.01															
140mm standard size blocks		10	.014															
150mm standard size blocks		10	.015															
190mm standard size blocks		10	.019															
215mm standard size blocks		10	.022															

*215mm thick walls can be built to show stretcher bond and give a face finish if required both sides if steel mesh bed joint reinforcement or butterfly wire ties are incorporated every 4th course.

†Garden wall bonds are intended for 215mm thick free standing boundary walls using 100% facing bricks.

Note: F = facings; C = commons; M = mortar

Index

Printed in the United Kingdom
by Lightning Source UK Ltd.
102818UKS00001B/49-510